Lecture Notes in Mathematics

Edited by A. Dold and B. Eckmann

1355

Jürgen Bokowski
Bernd Sturmfels

T0222326

Computational Synthetic Geometry

Springer-Verlag

Berlin Heidelberg New York London Paris Tokyo Hong Kong

Authors

Jürgen Bokowski
University of Darmstadt, Department of Mathematics
Schloßgartenstr. 7, 6100 Darmstadt, Federal Republic of Germany
e-mail: XBR1DB0N @ DDATHD21.bitnet

Bernd Sturmfels
Cornell University, Department of Mathematics
Ithaca, New York 14853, USA

Mathematics Subject Classification (1980): 05B35, 14M15, 51A20, 52A25, 68C20, 68E99

ISBN 3-540-50478-8 Springer-Verlag Berlin Heidelberg New York
ISBN 0-387-50478-8 Springer-Verlag New York Berlin Heidelberg

© Springer-Verlag Berlin Heidelberg 1989
Printed in Germany

Printing and binding: Druckhaus Beltz, Hemsbach/Bergstr.
2146/3140-543210 – Printed on acid-free paper

Preface

Computational synthetic geometry deals with methods for realizing abstract geometric objects in concrete vector spaces. This research monograph considers a large class of problems from convexity and discrete geometry including constructing polytopes from simplicial complexes, vector geometries from incidence structures and hyperplane arrangements from oriented matroids.

We show that algorithms for these constructions exist if and only if arbitrary polynomial equations are decidable with respect to the underlying field. For many special cases practical symbolic algorithms are presented and discussed. The methods developed are applied to obtain new mathematical results on polytopes, projective configurations and the combinatorics of Grassmann varieties. The necessary background knowledge is reviewed briefly. The text is accessible to students with graduate level background in mathematics, and it will serve professional geometers and computer scientists as an introduction and motivation for further research.

For several years both authors have had a fruitful and enjoyable collaboration which culminates in the present work. The first five chapters of this monograph are essentially Bernd Sturmfels' 1987 Ph.D. thesis which was supervised by Victor Klee at the University of Washington, Seattle. Chapter VI was written jointly by both authors, while the remaining chapters were written by Jürgen Bokowski in 1988.

The first author thanks all of his students who have helped to make this monograph possible. He thanks the Deutsche Forschungsgemeinschaft for support from 1985 to 1986. But above all he thanks his wife, Barbara, for her patient and encouraging support throughout all of the years that this work was in preparation.

The second author wishes to express his sincere appreciation to Victor Klee and Branko Grünbaum for their advice and encouragement. He thanks the Alfred P. Sloane Foundation who supported his work with a Doctoral Dissertation Fellowship. He is particularly grateful to Hyungsook for her warm presence.

Oberwolfach, January 1989

Jürgen Bokowski and Bernd Sturmfels

Table of Contents

Chapter I

PRELIMINARIES

Computational geometry is a rapidly growing young field on the border line of mathematics and computer science. Based on old mathematical foundations as well as the modern computational theories of algorithms and complexity, it has various applications in areas such as computer vision, robotics, statistics, artificial intelligence and molecular conformation [49].

Although for many problems solutions, which are mathematically elegant as well as useful for yielding fast algorithms, have been found, there is an important direction in computer-oriented geometry that has barely been touched so far. In order to solve many difficult problems in applied geometry, it is necessary to focus on what could roughly be described as *computer-aided geometric reasoning*. Beyond the geometric analysis of given discrete objects in Euclidean space, it is necessary to study and further develop methods for computational algebraic geometry, convexity, and automated theorem proving.

The present exposition aims to develop a few mathematical steps in this direction. Under the theme *Computational Synthetic Geometry* we shall discuss algorithmic aspects of certain fundamental realizability problems in discrete geometry. These questions originated in a purely mathematical setting, and the intrinsic motivation will be emphasized throughout this exposition. Still, the practical computational problems, which go along with our approach, are closely related to fascinating and important applications.

1.1. What is "computational synthetic geometry" ?

Many computer programs dealing with geometry proceed "analytically" in the following sense. The input is given in the form of coordinates representing some geometric object, and the program is supposed to analyze this object with respect to its geometric properties. These properties, mostly decoded numerically, then form the output of the program.

In *multidimensional sorting* [70] for example, the problem is to determine the order type of a finite point set $\{x_1, \ldots, x_n\}$ in real affine $(d-1)$-space. Here the *order type* is given by the orientations of all simplices spanned by ordered d-subsets $\{x_{\lambda_1}, x_{\lambda_2}, \ldots, x_{\lambda_d}\}$. In *scene analysis* [48] coordinates for certain planar projections of 3-dimensional objects are given, and it is the task of the computer program to reconstruct the original object. Also many problems from mathematical programming can be interpreted as a geometric analysis of a given set of input coordinates: linear optimization reduces to the feasibility problem of whether a certain convex polyhedron, given as an intersection of halfspaces, is empty or not ?

Most of the questions that have been studied so far in the young discipline of *computational geometry* fit into this framework. Both the problems and the solution procedures are essentially analytic in nature. Here are three typical examples from the fundamental book by Preparata & Shamos [123]: Given n points in the plane, find those whose mutual distance is smallest (Sect. 5.1). Given two convex d-polytopes P and Q, determine their intersection $P \cap Q$ (Sect. 7.3). Given a set S of n points in $\mathbf{R}^d$, identify those that are vertices of the convex hull $\text{conv}(S)$ (Sect. 3.2).

Though the above-mentioned problems are very different in their application, complexity and mathematical depth, they are all "analytic" in the sense mentioned earlier. In order to illustrate clearly the extent to which the geometric problems and algorithms to be discussed in this exposition are of a different flavor, let us now consider five classes of "synthetic" problems.

The algorithmic Steinitz problem

A *d-polytope* $P \subset \mathbf{R}^d$ is the convex hull of a finite affinely spanning subset of $\mathbf{R}^d$. The *face lattice* $\mathcal{F}(\mathcal{P})$ is the set of all faces of P ordered by inclusion and augmented by the empty set $\emptyset$ and P itself. Two polytopes P and P' are *combinatorially equivalent* if their face latties are isomorphic, and we call an arbitrary lattice $\mathcal{L}$ *polytopal* if $\mathcal{L} = \mathcal{F}(P)$ for some polytope P. For terminology and the basic properties of polytopes and their face lattices we refer to [76], [116].

The classification of all combinatorial types of polytopes of a given dimension and number of vertices has a long tradition in combinatorial convex geometry. For polytopes of dimension 3, the first non-trivial case, this problem is completely recuced to a combinatorial criterion by Steinitz' theorem [141], [60, Theorem 13.1]: *A graph G is isomorphic to the edge graph of a convex 3-polytope if and only if G is planar and 3-connected.* Using the language of simplicial topology [87] this statement is equivalent to: *Every piecewise linear 2-sphere is polytopal.* The same result has been proved for spheres with few vertices by Mani and Kleinschmidt [49, Theorem (15)]: *Every p.l. $(k-1)$-sphere with at most $k + 3$ vertices is polytopal.*

However, still only very little is known in the general case, and so far all attempts to solve the Steinitz problem, i.e. to find an intrinsic (purely combinatorial) characterization for face lattices of higher-dimensional polytopes, have been unsuccessful. In fact, results of G.Kalai [91] and B.Sturmfels [145], [146] show that in a certain local sense no such criterion exists. These results along with Mněv's universality theorem ([121], see Section 6.3) strongly support the algorithmical approach to the Steinitz problem.

Grünbaum & Sreedharan [82], Altshuler [3], [5], and other authors obtained complete classifications for 4-polytopes with few vertices by first enumerating all spheres and then deciding their polytopality. While a large class of polytopes could be constructed by geometric methods, see [6], these methods were not applicable in several "hard" cases. Subsequently, J.Bokowski introduced a combinatorial and algebraic reduction technique that led to solving these hard cases [5], [23],[24],[29] as well. His reduction of the Steinitz problem to oriented matroids motivated an intensive study of the realizability problem for oriented matroids and its applications to polytopes [30],[31],[32],[33], [34], and most of our investigations are based on the earlier work of Bokowski, see also Chapter 8 for historical remarks. From this perspective we tend to view the algorithmic Steinitz problem "*Given a lattice $\mathcal{L}$, is it polytopal ?*" as the first problem in computational synthetic geometry.

Projective incidence theorems

Given a field K, we say that two finite sets $X, Y \subset K^d$ are of the same *incidence type* if there is a bijection $\tau : X \to Y$ such that, for each $X' \subset X$, X' is linearly dependent if and only if $\tau(X')$ is linearly dependent. The case of *affine* rather than linear dependence can be reduced to the above by the usual embedding of affine $(d-1)$-space $A^{d-1}(K)$ as a hyperplane into K^d. The following natural question will be seen to be equivalent to the coordinatizability problem for matroids in Section 1.3. *Characterize or give an enumeration procedure for all incidence types over a given field K.*

By an *incidence theorem* we mean a statement which implies certain dependencies among points in affine space under the assumption of a given list of other incidences or dependencies. The theorems of Pappus and Desargues are probably the most important such theorems, and we shall employ the corresponding planar configurations as non-trivial examples in Chapters III and IV.

Let us briefly discuss a three-dimensional incidence theorem.

Proposition 1.1. (The Bundle Condition) *Given four lines in affine 3-space K^3, K any field, such that five of the six pairs of lines are coplanar, then also the sixth pair is coplanar.*

How can we design a "geometry theorem prover" that automatically proves such assertions or provides counterexamples ? Certainly, a counterexample would be given in terms of line coordinates, but what kind of output is acceptable as a proof ? Moreover, which formulation of Proposition 1.1 would be most suitable as input for such a program ?

We suggest that an "incidence conjecture" be considered as a realizability problem for possible counterexamples. Then there is either a realization or the structure is non-realizable in which case the theorem holds. So, instead of the bundle condition we would study a synthetic question of the following type.

Are there distinct $x_1, x_2, \ldots, x_8 \in K^3$ such that $\det(x_1, x_2, x_5, x_6) = \det(x_1, x_3, x_5, x_7) = \det(x_1, x_4, x_5, x_8) = \det(x_2, x_3, x_6, x_7) = \det(x_2, x_4, x_6, x_8) = 0$ but $\det(x_i, x_j, x_k, x_l) \neq 0$ for all other index quadruples (i, j, k, l) ?

A very short algebraic proof for the negative answer (and hence for Proposition 1.1) based on the method of final polynomials can be found in [33]. The following open problem in the same spirit has been posed by Boros, Füredi & Kelly [37], see also [92].

Problem 1.2. *Does there exist a finite set* $\mathcal{L}$ *of pairwise skew, affinely spanning lines in* $\mathbf{R}^4$ *such that the 3-space spanned by any two of them contains two additional lines from* $\mathcal{L}$ *?*

Arrangements of lines and pseudo-lines

A finite family $\mathcal{A}$ of (straight) lines in the real projective plane $P^2(\mathbf{R})$, not all of which pass through one point, constitutes an *arrangement of lines* [77]. If no point of $P^2(\mathbf{R})$ is contained in more than two lines, then the arrangement $\mathcal{A}$ is *simple*. $\mathcal{A}$ decomposes $P^2(\mathbf{R})$ into a two-dimensional cell complex, the 0-faces, 1-faces and 2-faces of which are called *vertices*, *edges* and *facets* of $\mathcal{A}$. Two arrangements are *isomorphic* if the corresponding cell complexes are isomorphic.

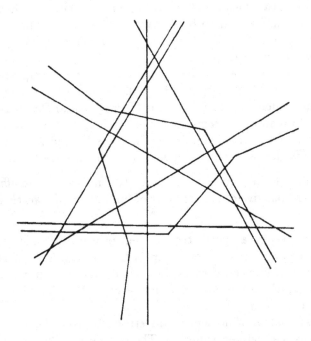

Figure 1-1. An arrangement of nine pseudo-lines which is not stretchable

In order to achieve a better understanding of the geometric and combinatorial properties of line arrangements, these notions have been generalized to arrangements in which "curved lines" are permitted. More precisely, an *arrangement of pseudo-lines* is a finite family $\mathcal{A}$ of simple closed curves in $P^2(\mathbf{R})$ such that any two curves have exactly one point in common, each crossing the other at this point, while no point is common to all curves.

5

An arrangement $\mathcal{A}$ of pseudo-lines is *stretchable* if it is isomorphic to an arrangement of (straight) lines. While every arrangement of $n \leq 8$ pseudo-lines is stretchable by a result of Goodman & Pollack [69], there exist non-stretchable arrangements with 9 pseudo-lines due to Ringel [126], see Figure 1-1. For 9 pseudo-lines this is essentially the only one according to J.Richter [125].

But in general there are large classes of non-strechable arrangements. A result in [30] states that for every $n \geq 5$ there is a non-stretchable simple arrangement Π_n of $2n$ pseudo-lines such that every proper subarrangement of Π_n is stretchable. This settles an old conjecture of Grünbaum [60, Conjecture 18.3].

The problem of enumerating all isomorphism types of arrangements with a given number of lines reduces to the purely combinatorial problem of enumerating all pseudo-line arrangements and the difficult geometric problem of deciding their stretchability. Here is the corresponding problem in computational synthetic geometry : *Find an algorithm that decides whether a given arrangement of pseudo-lines is stretchable.*

Diophantine problems in combinatorial geometry

Given a configuration of points and lines in the plane, can it be constructed with pencil and ruler alone ? The modern formulation of this question, which had occupied many geometers during the nineteenth century, asks whether a given (affine) incidence type over the real numbers can also be realized over the rational numbers. Figure 1-2 shows a well-known example of 9 points in the *real* plane such that no configuration of 9 points in the *rational* plane has the same incidence type, see [60, Chapter 5].

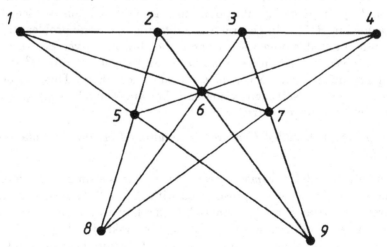

Figure 1-2. A non-rational incidence type in the real plane

In Chapter II we shall construct non-rational incidence types in the real plane related to arbitrary finite algebraic field extensions of the rationals. These and all other known non-rational incidence types share the property that some point of the configuration is contained in at least four (spanned) lines or some line is incident to at least four points. The following problem has been suggested by B. Grünbaum.

Conjecture 1.3. *Let I^K denote an incidence type of n points in the real plane with coordinates in the field K, such that no four points are one a line, and I^K has precisely n three point lines. Then I^K is necessarily realizable with rational coordinates.*

In Chapter III we shall discuss an algorithmic approach to this problem, and we shall prove Conjecture 1.3 for $n \leq 10$. In view of our partial results we venture the following stronger version of Conjecture 1.3.

Conjecture 1.4. *An n-element subset of the real plane can be approximated by rational point sets of the same incidence type I^K provided I^K satisfies the hypothesis of Conjecture 1.3. In other words, in this case the space of rational realizations is dense in the space of real realizations of I^K.*

The embedding of triangulated manifolds

Let $\Delta \subset 2^E$ be an r-dimensional simplicial complex on a finite set E. An *embedding* of Δ into Euclidean d-space is a mapping $i : E \rightarrow \mathbf{R}^d$ such that $\operatorname{conv} i(F) \cap \operatorname{conv} i(F') = \operatorname{conv} i(F \cap F')$ for all $F, F' \in \Delta$.

Clearly, an r-dimensional simplicial complex Δ cannot be embedded into $\mathbf{R}^{r-1}$, and by mapping E onto points in a general position, we see that Δ can be embedded in $\mathbf{R}^{2r+1}$. In general it is a difficult geometric problem to determine the minimum possible dimension for an embedding of a given Δ. Note, for example, that a given 1-dimensional complex Δ can be embedded in $\mathbf{R}^2$ if and only if the graph associated with Δ is planar, a result due to Fary. This is about the only case where a general satisfactory answer is known. The following open problem is of special interest in geometry.

Problem 1.5. [60, p.253] *Can every triangulated orientable 2-manifold be embedded in $\mathbf{R}^3$?*

Here a *triangulated (combinatorial) 2-manifold* is a simplicial complex Δ such that the topological space $|\Delta|$ associated with Δ is a 2-dimensional manifold. In fact, already for triangulated 2-manifolds with relatively few vertices it is extremely hard to find embeddings, and the work of Bokowski et. al. [18], [35], [22], [21] on this subject also shows the importance of the computational point of view in this area of research. It is generally expected that the answer to Problem 1.5 is negative for sufficiently large genus. Let us close our list of synthetic problems with a special case of Problem 1.5.

Conjecture 1.6. (Duke [61]) *Every triangulated torus can be embedded in* $\mathbf{R}^3$.

In abstracting the common features of these five fields of application, we suggest that we call a computational geometry problem *synthetic* if the input consists of a combinatorial or geometric condition and the output consists of either coordinates for such a geometric object in a vector space over a field (mostly the real algebraic numbers), or a proof that such an object is not realizable.

Observe that to each synthetic problem (P) there is a corresponding analytic problem (P'), the computational complexity of which is, in general, of polynomial order with respect to the size of the input. For example, for the Steinitz problem (P) *"Given a lattice $\mathcal{L}$, is it polytopal ?"*
the corresponding analytic problem
(P') *"Given points $x_1, x_2, \ldots, x_n \in \mathbf{R}^d$, determine the face lattice of their convex hull."*
can be solved in $O(n^2 + F \cdot \log n)$, where F denotes the size of the output. This is a result of R.Seidel [138]. Using McMullen's Upper Bound Theorem [115] we have $F \in O(n^{\lfloor d/2 \rfloor})$.

At first sight it may seem that (P) reduces to (P') via a *nondeterministic* oracle, and so if (P') has a certain deterministic time complexity, then (P) has the same complexity with respect to a non-deterministic computation model. This, however, is false mainly because of the essential additional requirement "... or a proof that such an object is not realizable". In fact, and we shall discuss this point in detail in Chapter II, there is, a priori, no reason why the synthetic problem (P) should be *decidable* at all, even if (P') can be solved by a linear time algorithm.

Let us close this section with the remark that there are, of course, important problems which, due to their partially analytic nature, cannot be called synthetic, but which are related to the above five problems because they lead in a similar way to arbitrarily "bad" polynomial systems. As an example let us consider the famous *piano movers problem* [137] in robotics. The input consists of coordinates for the starting position and the destination of the robot, as well as for the obstacles to be avoided, and it is the job of the computer program to decide whether start and end point are contained in the same path-connected component of the semi-algebraic variety of allowable positions.

1.2. Invariant theory and the Grassmann-Plücker relations

According to Felix Klein's *Erlanger Programm*, geometry deals with properties that are invariant under the action of some transformation group. The objects we discussed in the preceding section, such as face lattices, incidence theorems or order types, are invariant under all linear transformations, and therefore the classical invariant theory of linear groups provides a natural framework for our investigations.

Let K be a field and $n \geq d$ positive integers. The (ordered) n-element subsets of K^d can be identified with the K-vector space K^{nd} of $n \times d$-matrices $A = (a_{ij})$

where $a_{ij} \in K$. The group $G := \mathrm{SL}(K, d)$ of $d \times d$-matrices with determinant 1 acts on the n-element subsets of K^d by right multiplication on the matrices in K^{nd}. *How can we characterize all polynomial functions on K^{nd} which remain invariant under this action ?* More precisely, we would like to determine the invariant subring

$$K[x_{ij}]^G \quad := \quad \{ f \in K[x_{ij}] \mid f = f \circ T \text{ for all } T \in G \}$$

with respect to the induced action of G on $K[x_{ij}]$, the ring of polynomial functions on K^{nd}. Rather than considering all linear transformations, we restrict ourselves to the special linear group $G = \mathrm{SL}(K, d)$ for mainly technical reasons.

To give a geometric interpretation of the above question : *which properties of a vector configuration $\{x_1, \ldots, x_n\} \subset \mathbf{R}^d$ are invariant under linear transformations with determinant 1 ?* For every $\lambda := \{\lambda_1, \ldots, \lambda_d\} \subset \{1, 2, \ldots, n\}$ the volume of the parallelotope spanned by $x_{\lambda_1}, \ldots, x_{\lambda_d}$ is such an invariant magnitude. It is a consequence of the *First Fundamental Theorem of Invariant Theory* stated below that these volumes form a complete set of invariants for the action of $SL(\mathbf{R}, d)$ on n-vector configurations in $\mathbf{R}^d$. In other words, for all the synthetic problems that we suggested in Section 1.1, it is sufficient to perform the computations with these volumes as variables. To make this idea more precise let

$$\Lambda(n, d) \quad := \quad \{ [\lambda_1 \lambda_2 \ldots \lambda_d] \mid 1 \le \lambda_1 < \lambda_2 < \ldots < \lambda_d \le n \}$$

denote the $\binom{n}{d}$-element set of ordered d-tuples from $\{1, 2, \ldots, n\}$, and consider the polynomial ring $K[\Lambda(n, d)]$ which is freely generated over K by the set $\Lambda(n, d)$ of *brackets*. In this ring we abbreviate $[\lambda] := [\lambda_1 \lambda_2 \ldots \lambda_d]$ and

$$[\lambda_{\pi(1)} \lambda_{\pi(2)} \ldots \lambda_{\pi(d)}] \quad := \quad \mathrm{sign}\,(\pi) \cdot [\lambda]$$

for all permutations π. Since the determinant of a $d \times d$-matrix is invariant under the action (right multiplication) of G, the ring $K[x_{ij}]^G$ clearly contains the image of the ring homomorphism

$$\phi_{n,d} \;:\; K[\Lambda(n, d)] \quad \longrightarrow \quad K[x_{ij}]$$

$$[\lambda] \quad \longmapsto \quad \det \begin{pmatrix} x_{\lambda_1 1} & x_{\lambda_1 2} & \cdots & x_{\lambda_1 d} \\ x_{\lambda_2 1} & x_{\lambda_2 2} & \cdots & x_{\lambda_2 d} \\ \vdots & \vdots & \ddots & \vdots \\ x_{\lambda_d 1} & x_{\lambda_d 2} & \cdots & x_{\lambda_d d} \end{pmatrix}$$

In fact, the desired invariant ring is equal to the image of $\phi_{n,d}$. For a proof of this result see, e.g. [54], [70, Section VII.7], [138, Section II.A.6].

Theorem 1.7. (First Fundamental Theorem of Invariant Theory) *With the notations and hypotheses as above,* $\operatorname{Im} \phi_{n,d} = K[x_{ij}]^G$.

Theorem 1.7 implies that the invariant ring $K[x_{ij}]^G$ is isomorphic to the quotient $K[\Lambda(n,d)] / \operatorname{Ker} \phi_{n,d}$, and thus we need to find a description of the ideal $\operatorname{Ker} \phi_{n,d}$. Let us first see that this ideal can be described implicitly as the vanishing ideal of the Grassmann variety of simple d-vectors (= decomposable antisymmetric tensors) in the d-th exterior product $\wedge_d K^n$ of the vector space K^n. Here, $\Xi \in \wedge_d K^n$ is *simple* if and only if

$$\Xi = m_1 \wedge \ldots \wedge m_d \qquad \text{for some vectors } m_i \in K^n. \tag{1}$$

In fact, $K[\Lambda(n,d)]$ can be viewed as the coordinate ring of the $\binom{n}{d}$-dimensional K-vector space $\wedge_d K^n$, and $\phi_{n,d}$ is just the pullback of the map

$$\phi_{n,d}^* : K^{nd} \rightarrow \wedge_d K^n$$
$$(m_1, m_2, \ldots, m_d)^T \mapsto m_1 \wedge m_2 \wedge \ldots \wedge m_d.$$

In identifying $\Xi \in \wedge_d K^n$ with its Plücker coordinate map $\Xi : \Lambda(n,d) \to K$, we note that condition (1) is equivalent to the following :

There exists $M = (m_j^i) \in K^{nd}$ such that

$$\Xi(\lambda) = \Xi(\lambda_1, \ldots, \lambda_d) := \det \begin{pmatrix} m_1^{\lambda_1} & m_1^{\lambda_2} & \ldots & m_1^{\lambda_d} \\ m_2^{\lambda_1} & m_2^{\lambda_2} & \ldots & m_2^{\lambda_d} \\ \vdots & \vdots & \ddots & \vdots \\ m_d^{\lambda_1} & m_d^{\lambda_2} & \ldots & m_d^{\lambda_d} \end{pmatrix} \tag{2}$$

for all $\lambda \in \Lambda(n,d)$.

The set $G_{n,d}^K := \operatorname{Im} \phi_{n,d}^*$ of simple d-vectors is refered to as the *Grassmann variety* (it is Zariski closed by Theorem 1.8) since $\phi_{n,d}^*$ induces a natural topological isomorphism with the *Grassmann manifold* of d-dimensional vector subspaces of K^n. Not only did this geometric interpretation guide Grassmann himself (see Dieudonné [53]), but it is this subspace interpretation that is usually employed to prove the Grassmann-Plücker relations, see e.g. [38], [74].

The above geometric interpretation of simple d-vectors as (orbits under G of) n-element subsets of $\mathbf{R}^d$, however, yields a much more elementary proof for both the sufficiency and the necessity of the quadratic Grassmann-Plücker relations. Thus we suggest that Ξ be thought of as a mapping which assigns to each $\lambda \in \Lambda(n,d)$ the "oriented volume" of the parallelepiped spanned by the row vectors $m^{\lambda_1}, \ldots, m^{\lambda_d}$ in K^d. Of course, in talking about oriented volumes, we have the case in mind where K is a subfield of the real numbers. Still, the proof remains valid for an arbitrary field K.

Theorem 1.8. (Grassmann-Plücker relations) *A mapping* $\Xi : \Lambda(n,d) \to K$, K *any field, is simple if and only if it fulfils the relation*

$$\sum_{i=1}^{d+1} (-1)^i \cdot \Xi(\lambda_1, \ldots, \lambda_{i-1}, \lambda_{i+1}, \ldots, \lambda_{d+1}) \cdot \Xi(\lambda_i, \mu_1, \ldots, \mu_{d-1}) = 0 \qquad (3)$$

for all $\lambda \in \Lambda(n, d+1)$ *and* $\mu \in \Lambda(n, d-1)$.

From this we obtain, as an immediate consequence, an explicit description of the invariant ring $K[x_{ij}]^G$ provided the field K is infinite. Let $I_{n,d}^K$ denote the ideal in $K[\Lambda(n,d)]$ which is generated by the *Grassmann-Plücker syzygies*

$$\sum_{i=1}^{d+1} (-1)^i \cdot [\lambda_1 \ldots \lambda_{i-1}\lambda_{i+1} \ldots \lambda_{d+1}] \cdot [\lambda_i \mu_1 \ldots \mu_{d-1}]$$

for all $\lambda \in \Lambda(n, d+1)$ and $\mu \in \Lambda(n, d-1)$. By Theorem 1.8, the syzygy ideal $I_{n,d}^K$ is the vanishing ideal of the set $\operatorname{Im} \phi_{n,d}^*$. Since the ring homomorphism $\phi_{n,d}$ is the pullback of $\phi_{n,d}^*$, that is $\phi_{n,d}(p) = p \circ \phi_{n,d}^*$ for all $p \in K[\Lambda(n,d)]$, $I_{n,d}^K$ is precisely the kernel of $\phi_{n,d}$ by [82, Exercise I.2.3] provided K is infinite. This proves

Corollary 1.9. *Given an infinite field* K, *then the invariant ring* $K[x_{ij}]^G$ *is isomorphic to the coordinate ring* $K[G_{n,d}]$ *of the Grassmann variety* $G_{n,d}^K$ *of simple d-vectors in* $\wedge_d K^n$.

Proof of Theorem 1.8. Assuming Ξ to be simple, there exist (row) vectors $m^1, m^2 \ldots, m^n \in K^d$ such that $\Xi(\sigma_1, \ldots, \sigma_d) = \det(m^{\sigma_1}, \ldots, m^{\sigma_d})$ for all $\sigma \in \Lambda(n,d)$.

Given $\lambda \in \Lambda(n, d+1)$ and $\mu \in \Lambda(n, d-1)$, we may suppose that the sum in (3) has at least one non-vanishing term. Then the vectors $m^{\lambda_1}, \ldots, m^{\lambda_{d+1}}$ span K^d and hence fulfil a unique dependency (up to scaling) which follows immediately from Cramer's rule.

$$\sum_{i=1}^{d+1} (-1)^i \cdot \det(m^{\lambda_1}, \ldots, m^{\lambda_{i-1}}, m^{\lambda_{i+1}}, \ldots, m^{\lambda_{d+1}}) \cdot m^{\lambda_i} = 0 \qquad (4)$$

In plugging the representation of the zero vector in (4) into the first slot of the vanishing determinant $\det(0, m^{\mu_1}, \ldots, m^{\mu_{d-1}})$, we obtain the desired result by the linearity of the determinant function in the first slot

$$\det\Big(\sum_{i=1}^{d+1} (-1)^i \cdot \det(m^{\lambda_1}, \ldots, m^{\lambda_{i-1}}, m^{\lambda_{i+1}}, \ldots, m^{\lambda_{d+1}}) \cdot m^{\lambda_i},$$

$$m^{\mu_1}, \ldots, m^{\mu_{d-1}} \Big) \qquad (5)$$

$$= \sum_{i=1}^{d+1} (-1)^i \cdot \det(m^{\lambda_1},\ldots,m^{\lambda_{i-1}},m^{\lambda_{i+1}},\ldots,m^{\lambda_{d+1}}) \cdot \det(m^{\lambda_i},m^{\mu_1},\ldots,m^{\mu_{d-1}})$$

$$= \sum_{i=1}^{d+1} (-1)^i \cdot \Xi(\lambda_1,\ldots,\lambda_{i-1},\lambda_{i+1},\ldots,\lambda_{d+1}) \cdot \Xi(\lambda_i,\mu_1,\ldots,\mu_{d-1}) \quad = \quad 0$$

Conversely, given that Ξ satisfies the Grassmann-Plücker relations (3) we shall exhibit an $n \times d$-matrix with row vectors $m_1,\ldots,m_n \in K^d$ such that (2) holds. If Ξ is zero, we can simply choose zero vectors; otherwise it can be assumed w.l.o.g that

$$\Xi(1,2,\ldots,d) \quad \neq \quad 0. \tag{6}$$

Choose any basis $(m^1,\ldots,m^d)$ of K^d such that

$$\det(m^1,m^2,\ldots,m^d) \quad = \quad \Xi(1,2,\ldots,d), \tag{7}$$

and define

$$m^j \quad := \quad \frac{1}{\Xi(1,2,\ldots,d)} \sum_{i=1}^{d} \Xi(1,\ldots,i-1,j,i+1,\ldots,d) \cdot m^i \tag{8}$$

for $j = d+1,\ldots,n$. We prove by induction on k that (2) holds for all d-tuples $\lambda \in \Lambda(k,d)$ where $d \leq k \leq n$. For $k = d$ this is guaranteed by our choice in (7).

Now, let $k > d$ and assume that (2) holds for all $\lambda \in \Lambda(k-1,d)$. Let $\lambda \in \Lambda(k,d) \setminus \Lambda(k-1,d)$, i.e. $\lambda_d = k$. Then

$$\det(m^{\lambda_1},\ldots,m^{\lambda_{d-1}},m^k)$$

$$= \quad \det\left(m^{\lambda_1},\ldots,m^{\lambda_{d-1}}, \left\{ \frac{1}{\Xi(1,2,\ldots,d)} \sum_{i=1}^{d} \Xi(1,\ldots,i-1,k,i+1,\ldots,d)m^i \right\} \right)$$

$$= \quad \frac{1}{\Xi(1,\ldots,d)} \sum_{i=1}^{d} \det(m^{\lambda_1},\ldots,m^{\lambda_{d-1}},m^i)\, \Xi(1\ldots,i-1,k,i+1,\ldots,d) \tag{9}$$

$$= \quad \frac{1}{\Xi(1,\ldots,d)} \sum_{i=1}^{d} \Xi(\lambda_1,\ldots,\lambda_{d-1},i)\, \Xi(1\ldots,i-1,k,i+1,\ldots,d) \tag{10}$$

$$= \quad \Xi(\lambda_1,\ldots,\lambda_{d-1},k).$$

In the step from (9) to (10) we use the induction hypothesis because the indices $\lambda_1,\ldots,\lambda_{d-1}$ and i are all smaller than k. The last equation is just one of the Grassmann-Plücker relations, which, by assumption, holds for the d-vector Ξ. $\quad\square$

1.3. Matroids and oriented matroids

The realizability problem for matroids and oriented matroids are computational synthetic geometry problems of fundamental and paradigmatic character. Many problems, in particular all the questions mentioned in Section 1.1, can be reduced to these two problems in a theoretically (and sometimes also algorithmically) straightforward manner. Other synthetic problems which cannot be simply reduced to these two, like problems from *Distance Geometry, Rigidity* or *Symmetric Realizability,* are closely related, and our methods and results are likely to be helpful for such applications as well.

Realizable matroids and realizable oriented matroids are equivalence classes of finite subsets of K^d corresponding to precisely the incidence types and order types as defined above. Since incidence and order types are invariant under the canonical action of the linear group on K^d, (realizable) matroids and oriented matroids can also be thought of as the cells in natural partitions of the Grassmann manifold $G_{n,d}^K$. As is pointed out in an article by Gelfand et.al. [68], the study of the stratification of $G_{n,d}^K$ induced by this matroid partition is related to interesting questions in various mathematical subfields, compare Section 4.4.

Matroids and oriented matroids are combinatorial structures generalizing these (realizable) incidence types and order types. On the one hand, it is difficult (or impossible) to give an easy intrinsic description for all realizable structures over a given field K, and consequently one studies "locally realizable structures" instead. On the other hand, one is interested in working in the broadest possible setting in which certain familiar combinatorial properties of linear algebra still hold. One basic property, the Steinitz exchange lemma, gives rise to the following definition of matroids.

Definition 1.10. *A matroid M is a pair $(E, \mathcal{B})$ where E is finite set and $\mathcal{B}$ a collection of subsets of E, called the bases of M, such that for all $B, B' \in \mathcal{B}$ and $e \in B \setminus B'$ there exists $e' \in B' \setminus B$ such that $B \setminus \{e\} \cup \{e'\} \in \mathcal{B}$.*

The finite set E will frequently be identified with $\{1, 2, \ldots, n\}$ throughout this text. It follows directly from Definition 1.10 that all bases of M have the same cardinality d, called the *rank* of M. Given a field K, M is *realizable* over K (the terms *coordinatizable, linear* and *representable* are used synonymously throughout the literature) if there exists a map $\sigma : E \to K^d$ such that $B \subset E$ is a basis of M if and only if $\sigma(B)$ is a basis of K^d. The set $X = \sigma(E) \subset K^d$ (or the corresponding coordinate matrix) is a *realization* or *coordinatization* of M. It is a *weak coordinatization* of M if $B \subset E$ is a basis of M whenever $\sigma(B)$ is a basis of K^d (but not necessarily vice versa).

Two subsets $X = \{x_1, x_2, \ldots, x_n\}$ and $Y = \{y_1, y_2, \ldots, y_n\}$ of K^d are *projectively equivalent* if there exists a linear transformation $A \in GL(K, d)$ and $\mu_1, \ldots, \mu_n \in K \setminus \{0\}$ such that $y_i = \mu_i A x_i$ for $i = 1, \ldots, n$. Clearly, projectively equivalent sets have the same incidence type, i.e. they realize the same matroid. A matroid M is *projectively* unique if any two realizations of M are projectively equivalent.

Flexibility in the underlying axiomatics is one of the reasons why matroid theory plays such an important role in many parts of discrete mathematics. Apart from Definition 2.1, there exist at least twenty different axiomatizations for matroids, or *combinatorial geometries* as they are sometimes called, which at first glance seem to have little in common. So, a matroid can be considered as geometric lattice, hull operator, collection of bases, simplicial complex of independent sets, set of circuits, closed flats ... etc. Without going into any details or proofs, let us briefly sketch a few basics of matroid theory. As introductory texts to matroid theory, we recommend [96], [158], [161].

Let M be a matroid on an n-element set E as defined above. $A \subset E$ is *independent* if A is contained in some basis of M, and *dependent* otherwise. The *rank function* ρ assigns to every subset $A \subset E$ the maximal cardinality $\rho(A)$ of an independent subset contained in A. A subset $A \subset E$ is a *(closed) flat* of M if $\rho(A \cup x) > \rho(A)$ for all $x \in E \setminus A$. Flats of rank $1, 2, 3$ are called *points, lines, planes*, and flats of rank $d - 1$ are called *hyperplanes* of M.

These definitions indicate that starting with a single axiom for defining matroids, a rich geometric theory can be developed. We will not pursue that idea further here because it is our main concern to find coordinates for a given matroid M or prove that M is not realizable. In one point, however, we have to be careful: the question whether a matroid is realizable depends heavily on the field we are interested in as we shall see in various examples in the following sections.

In the cases of *binary* ($K = \mathbf{GF}_2$), *ternary* ($K = \mathbf{GF}_3$) and *regular* matroids (realizable over every field), it is possible to characterize realizable matroids by the exclusion of finite collections of "forbidden minors". These characterizations due to Tutte and Seymour are generally viewed as the deepest classical results on the coordinatizability of matroids [96]. See Trümper [155] for an efficient algorithm with regard to the question of whether a given matroid is regular. From our point of view, however, it is the case char $K = 0$ which is most interesting, and according to a famous theorem of Vamos [156], a forbidden minor characterization cannot exist in that case.

The dual M^* of a rank d matroid M on E is a rank $n - d$ matroid on E, whose bases are precisely the set complements of the bases of M. Given $e \in E$, the *deletion* $M \setminus e$ is the rank d matroid on $E \setminus \{e\}$ consisting of all bases of M which do not contain e. The *contraction* M/e is the rank $d - 1$ matroid on $E \setminus \{e\}$, whose bases are the sets B for which $B \cup \{c\}$ is a basis of M. A *minor* of M is any matroid M' obtainable from M by repeated application of the deletion and contraction operation. In this case M is an *extension* of M'. A matroid is *simple* provided, for each $A \subset E$, $\rho(A) = 1$ if and only if A is a singleton. In practically all of the questions being discussed in this paper, we can restrict ourselves to simple matroids. As it is easy to see that by deleting redundant elements of E (corresponding to *loops* and *parallel elements*), every matroid M has a unique maximal simple minor $\widetilde{M}$ which still carries the entire geometric structure of M.

Given another rank d matroid M' on E, we say that M' is a *weak image* of M, abbreviated $M' \leq M$, provided every basis of M' is a basis of M.

There is a very close connection between matroid theory and Grassmann algebra [55], [95] of which we will make use whenever appropriate. This aspect, which can also be helpful to gain a deeper understanding of duality theory and decompositions of matroids, leads in particular to the following alternative view of matroids. A rank d matroid M on $E = \{1, 2, \ldots, n\}$ can be seen as a mapping $M : \Lambda(n, d) \to \{0, 1\}$ where $M(\lambda) = 1$ if and only if the set λ is a basis of M. Then M is K-realizable if and only if there exists a simple $\Xi \in G_{n,d}^K \subset \wedge_d K^n$ with $M = \text{supp} \, \Xi$, where

$$\begin{aligned} \text{supp} \quad : \quad K \quad &\to \quad \{0, 1\} \\ x \quad &\mapsto \quad \begin{cases} 0 & \text{if } x = 0 \\ 1 & \text{otherwise} \end{cases} \end{aligned}$$

Consider the following commutative addition and multiplication rules within the set $\{0, 1, *\}$.

$$0 \cdot 0 = 0 \cdot 1 = 0 \cdot * = 0, \quad 1 \cdot 1 = 1, \quad 1 \cdot * = * \cdot * = *$$

$$0 + 0 = 0, \quad 0 + 1 = 1 + 1 = 1, \quad 0 + * = 1 + * = * + * = *$$

Here "0" and "1" stand for the zero and non-zero elements of a field and "$*$" stands for "don't know". With these operations, $\{0, 1, *\}$ has the structure of a *fuzzy ring*. Fuzzy rings have been employed by A. Dress to develop his theory of *matroids with coefficients* which generalize matroids and oriented matroids as well as incidence types over fields [58].

Remark 1.11. $M : \Lambda(n, d) \to \{0, 1\}$ *is a matroid if and only if*

$$\sum_{i=1}^{d+1} M(\lambda_1, \ldots, \lambda_{i-1}, \lambda_{i+1}, \ldots, \lambda_{d+1}) \cdot M(\lambda_i, \mu_1, \ldots, \mu_{d-1}) \quad \subset \quad \{0, *\}$$

in the fuzzy ring $\{0, 1, *\}$ *for all* $\lambda \in \Lambda(n, d+1)$ *and* $\mu \in \Lambda(n, d-1)$.

This definition of a matroid based on the Grassmann-Plücker relations, see Theorem 1.8, leads naturally to oriented matroids where the non-zero elements represented by "1" are seperated into positive and negative elements. More precisely, we consider the fuzzy ring $\{-1, 0, +1, *\}$ with the commutative multiplication and addition

$$0 \cdot 0 = 1 \cdot 0 = 0 \cdot (-1) = 0 \cdot * = 0, \quad 1 \cdot 1 = (-1) \cdot (-1) = 1, \quad 1 \cdot (-1) = -1,$$

$$(+1) + (-1) = 1 \cdot * = (-1) \cdot * = * \cdot * = *, \quad 0 + 0 = 0, \quad 0 + 1 = 1 + 1 = 1,$$

$$0 + (-1) = (-1) + (-1) = -1, \quad * + 1 = * + (-1) = * + 0 = * + * = *.$$

Definition 1.12. *A partial oriented matroid is a map*
$\chi : \Lambda(n,d) \rightarrow \{-1,0,+1,*\}$ *such that*

$$\sum_{i=1}^{d+1}(-1)^i \cdot \chi(\lambda_1,\ldots,\lambda_{i-1},\lambda_{i+1},\ldots,\lambda_{d+1}) \cdot \chi(\lambda_i,\mu_1,\ldots,\mu_{d-1}) \quad \subset \quad \{0,*\}$$

in the fuzzy ring $\{-1,0,+1,*\}$ *for all* $\lambda \in \Lambda(n,d+1)$ *and* $\mu \in \Lambda(n,d-1)$. χ *is an oriented matroid if* $\mathrm{Im}(\chi) \subset \{-1,0,+1\}$. *If* $\mathrm{Im}(\chi) \subset \{-1,+1\}$, *the oriented matroid* χ *is called uniform.*

There is a natural poset on the set of partial oriented matroids with fixed rank d and number of points n. χ_1 is said to be *above* χ_2 (resp. χ_2 is *below* χ_1) if $\chi_2(\lambda) \in \{\chi_1(\lambda),*\}$ for all $\lambda \in \Lambda(n,d)$. Given any finite ordered set E, we talk (as in the matroid case) of (partial) oriented matroids *on* E via the identification of E with $\{1,2,\ldots,n\}$ for some n.

An oriented matroid as defined above is also refered to as a *chirotope*, a term which has been coined by A.Dress and A.Dreiding for a combinatorial structure that reflects the geometric aspects of chirality in organic chemistry [56], [57]. It turned out that their chirotopes are equivalent to the oriented matroids which were axiomatized independently by several authors, see Las Vergnas [102], Bland & Las Vergnas [17] and Folkman & Lawrence [64], Buchi [40].

At the same time, also independently, J. Bokowski began to initiate a similar approach from a geometric point of view. His main interest was the Steinitz problem, and he developed a method based on chirotopes for deciding the polytopality of a given sphere, which was sucessfully applied in solving several research problems [5], [23],[29],[32].

The classical definition for oriented matroids in terms of signed circuits, given in [17], reads as follows: For a *signed vector* $X \in \{-1,0,+1\}^n$, we denote $X^+ := \{i \in \{1,\ldots,n\} \,|\, X_i = +1\}$. The sets X^0 and X^- are defined accordingly. A set $\mathcal{O} \subset \{-1,0,+1\}^n$ is called an *oriented matroid* if the following two conditions are fulfilled:

(0) For all $X \in \mathcal{O}$ we have $X \neq 0$ and $-X \in \mathcal{O}$; and for all $X_1, X_2 \in \mathcal{O}$, the inclusion $X_1^0 \subset X_2^0$ implies $X_1 = \pm X_2$.

(1) For all $X_1, X_2 \in \mathcal{O}$ such that $X_1 \neq X_2$, and all $i \in (X_1^+ \cap X_2^-) \cup (X_1^- \cap X_2^+)$, there exists $X_3 \in \mathcal{O}$ such that $X_3^+ \subset (X_1^+ \cup X_2^+) \setminus i$, $X_3^- \subset (X_1^- \cup X_2^-) \setminus i$.

The elements of $\mathcal{O}$ are called *signed circuits* of the oriented matroid $\mathcal{O}$. Given a d-dimensional vector subspace $\mathbf{V}$ of K^n, K an ordered field, the set of signed vectors of the *elementary vectors* of $\mathbf{V}$, that is non-zero vectors with minimal support, forms a K-*realizable* oriented matroid. In [100] Las Vergnas introduced convexity to oriented matroids, and several subsequent papers of other authors have proceeded to generalize many fundamental ideas, concepts and results on convex polytopes to the broader setting of oriented matroids, see e.g. Billera & Munson [15], Mandel [109], Cordovil & Duchet [45],[46] and Bachem [9].

For an equivalence proof of the two definitions see Lawrence [104]. Given an oriented matroid χ as in Definition 1.11, the *cocircuits* of χ are the signed vectors

$$\pm C_\mu(\chi) := \pm \begin{pmatrix} \chi(\mu_1,\ldots,\mu_{d-1},1) \\ \vdots \\ \chi(\mu_1,\ldots,\mu_{d-1},i) \\ \vdots \\ \chi(\mu_1,\ldots,\mu_{d-1},n) \end{pmatrix} \in \{-1,0,+1\}^n \quad \text{for } \mu \in \Lambda(n,d-1).$$

A partial oriented matroid χ is *realizable* if there exist $x_1,\ldots,x_n \in K^d$ such that

$$\chi(\lambda) = \text{sign}(\det(x_{\lambda_1},\ldots,x_{\lambda_d})) \quad \text{for all } \lambda \in \Lambda(n,d) \text{ with } \chi(\lambda) \neq *.$$

Given $X = \{x_1,\ldots,x_n\} \subset K^d$, the unique oriented matroid χ with the above property is called the *oriented matroid of linear dependencies* on X or the *linear oriented matroid* of X. The *affine oriented matroid* of $\{y_1,\ldots,y_n\} \subset K^{d-1}$ is the linear oriented matroid of $\{(1,y_1),\ldots,(1,y_n)\} \subset K^d$. The following easy observation states that oriented matroids can be interpreted as "locally realizable" sign patterns.

Remark 1.13. *Let* $\chi : \Lambda(n,d) \to \{-1,0,+1\}$ *such that* $M := \text{supp}\,\chi$ *is a matroid and such that every rank d minor of M with $d+2$ elements can be realized compatibly with the signs in* χ. *Then* χ *is an oriented matroid.*

The *dual oriented matroid* χ^* of a rank d oriented matroid with n vertices is defined as

$$\chi^* : \Lambda(n,n-d) \quad \to \quad \{-1,0,+1\}$$

$$\lambda \quad \mapsto \quad \chi(\{1,\ldots,n\} \setminus \lambda) * (-1)^{\sum\limits_{i=1}^{n-d} \lambda_i}$$

Observe that the above sign factor, the parity of $\sum\limits_{i=1}^{n-d} \lambda_i$, gives us the sign of the permutation (λ,λ^c), where $\lambda^c := \{1,\ldots,n\} \setminus \lambda$. A *circuit* of an oriented matroid χ is a cocircuit of its dual.

For a subset $A \subset \{1,\ldots,n\}$, the *reorientation* on A of χ, denoted $_{\overline{A}}\chi$, is defined by

$$_{\overline{A}}\chi(\lambda) \quad := \quad \chi(\lambda) \cdot (-1)^{|\lambda \cap A|}.$$

The set of all oriented matroids obtained from χ by reorientation is the *reorientation class* of χ. It is an important result due to Folkman & Lawrence [64] that the reorientation classes of rank d oriented matroids are in one-to-one correspondence with the (isomorphism types of) arrangements of pseudo-hyperplanes in projective $(d-1)$-space $P^{d-1}(K)$, where the prefix "pseudo" can be dropped if we require *realizable* oriented matroids.

We define the oriented matroids obtained from χ by *deletion*

$$\chi \setminus n := \chi \mid_{\Lambda(n-1,d)}$$

and by *contraction*

$$\chi/n := (\chi^* \setminus n)^*$$

for the vertex n. Relabeling with respect to alternation gives us the general case. An oriented matroid χ' is a *minor* of χ if it can be obtained from χ by repeated deletions and contractions in which case χ is an *extension* of χ'.

Given a k-polytope $P = \text{conv}\{y_1, \ldots, y_n\} \subset K^k$, we assign to P the affine rank $(k + 1)$ oriented matroid χ of $\{y_1, \ldots, y_n\}$. If $F = \text{conv}\{y_{i_1}, \ldots, y_{i_l}\}$ is a facet of P, then all vertices y_j of P which are not in F lie on the same side of the hyperplane affinely spanned by F. This is equivalent to the existence of a positive cocircuit $C = C^+$ with $C^0 = \{i_1, \ldots, i_l\}$.

Hence the following definition generalizes the concept of face-lattices of convex polytopes : $F \subset \{1, \ldots, n\}$ is called a *facet* of the oriented matroid χ if there is a positive cocircuit of χ with $C^0 = F$. Finite intersections of facets are called *faces* of χ. The set of all faces of χ, ordered by inclusion, forms the *face-lattice* $\text{FL}(\chi)$ of χ. A lattice S is a *matroid sphere* if $S = \text{FL}(\chi)$ for an oriented matroid χ. Clearly, a lattice S is polytopal if and only if there exists a *realizable* oriented matroid χ with $\text{FL}(\chi) = S$.

In the discussion of face lattices of oriented matroids we can restrict ourselves to *matroid polytopes*, that is to acyclic oriented matroids without interior points. An oriented matroid χ on E is *acyclic* if it has no positive circuit, and $e \in E$ is an *interior* point of χ if there is a circuit C of χ with $C^+ = \{e\}$. Since in the realizable case, signed circuits correspond to minimal Radon partitions, this combinatorial definition of interior points agrees with the usual one by Caratheodory's theorem [115]. Our arguments mentioned above imply in particular that if two polytopes P and $\tilde{P}$ have the same matroid polytope, then they are combinatorially equivalent. An octahedron with vertices in general position and a regular octahedron show that in general the converse fails to hold.

Chapter II

ON THE EXISTENCE OF ALGORITHMS

In this chapter we investigate the existence of algorithms for the realizability problems of matroids and oriented matroids as well as the algorithmic Steinitz problem over a fixed field K. We prove that such algorithms exist if and only if there exists a decision procedure for polynomial equations with integer coefficients in K. The latter question for the case $K = \mathbf{Q}$, the rational numbers, is the unsolved rational version of Hilbert's 10th problem, which, therefore, is equivalent to various Diophantine problems in computational synthetic geometry.

As is customary in the theory of computation, we state all algorithms in a brief verbal form but still detailed enough to be coded in some computer language if necessary. This attitude towards algorithms is justified by the *Church Thesis* which states that the intuitive notion of a "decision procedure" agrees with the standard computation models such as Turing machines, recursive functions or any higher programming language (see e.g. [105]).

2.1. Encoding affine algebraic varieties in matroids

It is well known in classical synthetic geometry that the algebraic operations of addition and multiplication have their geometric analogues in certain projectively invariant constructions which are commonly used for the coordinatization of Desarguesian projective planes, see e.g. [70, Sect. VI.7. "The algebra of points on a line"]. It is one of the earliest results in matroid theory due to MacLane [114] that with these constructions arbitrary polynomial conditions can be imposed on a variable point on a line in a rank 3 matroid. A nice proof for a slightly generalized version of MacLane's theorem has been given by N.White in [141, Section 7].

Theorem 2.1. (MacLane,White) *Let K be any finite algebraic extension field of a prime field K', that is K' is the field of rationals or $\mathbf{GF_p}$ for some prime p. Then there exists a rank 3 matroid M which is K-coordinatizable, such that if M is L-coordinatizable for some field extension L of K', then L contains a subfield isomorphic to K.*

In the case of rational numbers this result says that all algebraic numbers are necessary in order to coordinatize all rank 3 matroids. Or still in other words : arbitrary algebraic numbers can be encoded in suitable rank 3 matroids. See also the papers of Mnëv [119] and Sturmfels [145] where Theorem 2.1 is applied in contracting irrational convex polytopes related to arbitrary real algebraic numbers.

Here we further generalize Theorem 2.1 by showing that arbitrary affine algebraic varieties defined over some prime field K' can be encoded in rank 3 matroids.

More precisely, every such variety is birationally isomorphic to the (projective) realization space of some rank 3 matroid, compare Section 4.4. Although the "projective addition" and the "projective multiplication" are very well-known techniques in areas far removed from matroid theory, nobody seems to have noticed the computational significance of these constructions so far . It is the objective of this section to prove the following result.

Theorem 2.2. *For any field K, the following statements are equivalent.*
(1) There exists an algorithm for deciding any finite set of polynomials
   $f_1, \ldots, f_r \in \mathbf{Z}[x_1, \ldots, x_n]$, whether the f_i have a common zero in K^n.
(2) There exists an algorithm to decide for an arbitrary rank 3 matroid M whether
   M is coordinatizable over K.

The hard part of the theorem is, of course, the implication $(2) \Rightarrow (1)$. Although the underlying geometric construction is a straightforward extension of the standard arguments, we shall provide proofs that are sufficiently detailed to make sure that every step can indeed be carried out by a computing machine. Observe that, writing K' for the prime field of K, $\mathbf{Z}[x_1, \ldots, x_n]$ can be replaced in the above statement by $K'[x_1, \ldots, x_n]$. Every polynomial identity with integer coefficients is equivalent over K to its image in $K'[x_1, \ldots, x_n]$ and vice versa.

Proof of Theorem 2.2.
We first prove the implication $(1) \Rightarrow (2)$. Let M be a rank 3 matroid on a finite set E with the set of bases $\mathcal{B} \subset \binom{E}{3}$. We introduce three variables x_e, y_e and z_e for each $e \in E$ and a variable D_b for each $b \in \mathcal{B}$. Consider the following system of equations

$$\det \begin{pmatrix} x_{e_1} & y_{e_1} & z_{e_1} \\ x_{e_2} & y_{e_2} & z_{e_2} \\ x_{e_3} & y_{e_3} & z_{e_3} \end{pmatrix} = 0 \qquad \text{for all } \{e_1, e_2, e_3\} \in \binom{E}{3} \setminus \mathcal{B}$$

$$D_{\{e_1, e_2, e_3\}} \cdot \det \begin{pmatrix} x_{e_1} & y_{e_1} & z_{e_1} \\ x_{e_2} & y_{e_2} & z_{e_2} \\ x_{e_3} & y_{e_3} & z_{e_3} \end{pmatrix} - 1 = 0 \qquad \text{for all } \{e_1, e_2, e_3\} \in \mathcal{B}$$

defined over the polynomial ring $\mathbf{Z}[\{x_e, y_e, z_e, D_b : e \in E, b \in \mathcal{B}\}]$. Observe that the second class of equations corresponds to the fraction S_M^{-1} to be introduced in Chapter IV.

It follows from the definitions in Section 1.3 that this system has a solution in K if and only if M is K-coordinatizable.

In order to prove the second implication, it is our goal to describe an algorithm that assigns to every finite subset $\mathcal{F} = \{f_1, \ldots, f_r\} \subset \mathbf{Z}[x_1, \ldots, x_n]$ a family $\mathcal{M}_{\mathcal{F}}$ of rank 3 matroids such that some member of $\mathcal{M}_{\mathcal{F}}$ is K-coordinatizable if and only if the affine algebraic K–variety $V(\mathcal{F}) := \{a \in K^n : f(a) = 0 \text{ for all } f \in \mathcal{F}\}$

associated with $\mathcal{F}$ is non-empty. For simplicity let us first assume that $\mathcal{F} = \{p\}$ is a singleton.

We define a formal language L_n of *polynomial expressions* over the alphabet $\Sigma_n := \{x_1, x_2, \ldots, x_n, +, \cdot, (,), 1, -1\}$ as follows:

- $1, -1, x_1, x_2, \ldots, x_n$ are contained in L_n.
- If p and q are in L_n, so is $(p + q)$.
- If p and q are in L_n, so is $(p \cdot q)$.
- No other words are contained in L_n.

There is a canonical surjection $\sigma : L_n \to \mathbf{Z}[x_1, \ldots, x_n]$ which assigns to every polynomial expression the corresponding polynomial function. In using some normal form procedure for integer polynomials, we obtain a computable section $\nu : \mathbf{Z}[x_1, \ldots, x_n] \mapsto L_n$, i.e. $\sigma(\nu(f)) = f$ for all $f \in \mathbf{Z}[x_1, \ldots x_n]$. Thus we can replace $\mathbf{Z}[x_1, \ldots, x_n]$ in the statement of Theorem 2.2 by L_n.

Let K' denote the prime field of K. Again we have a canonical surjection $\sigma_{K'} : L_n \to K'[x_1, \ldots, x_n]$. Since there also exist normal form algorithms for polynomials with rational coefficients as well as polynomials over $\mathbf{GF_p}$, $\sigma_{K'}$ is a computable function. In particular, it is decidable whether for some $p \in L_n$ its image $\sigma_{K'}(p)$ is zero in $K'[x_1, \ldots, x_n]$ or in its field of fractions $K'(x_1, \ldots, x_n)$.

On the basis of this set-up the proof proceeds in three steps. First we describe a recursive procedure that assigns to every polynomial expression $p \in L_n$ a certain $3 \times N$ matrix A_p with entries in $L_n \cup \{0\}$. The recursive step corresponds to the classical projective addition–multiplication technique, see Figures 2-1 and 2-2. In a second step we interpret every such matrix A_p via $\sigma_{K'}$ as a matrix with entries from the field $K'(x_1, \ldots, x_n)$, and we compute the corresponding rank 3 matroid M_p. Finally, choosing suitable weak images of M_p we will obtain the desired family $\mathcal{M}_p$.

The 3×9-matrix

$$
A \;=\; \begin{pmatrix} 1 & 0 & 0 & 1 & 0 & 1 & 1 & 0 & -1 \\ 0 & 1 & 0 & 1 & 1 & 0 & 1 & 1 & 0 \\ 0 & 0 & 1 & 1 & 1 & 1 & 0 & -1 & 1 \end{pmatrix}
$$

plays the role of a reference matrix or *frame matrix*. Define for every $p \in L_n$ a matrix A_p as follows:

- $A_1 := A$ and $A_{-1} := A$.
- A_{x_i} is the matrix A with an additional 10th column $(1, 0, x_i)^T$.
- $A_{(p+q)} := (A_p, A_q, B)$ where A_p and A_q are the matrices corresponding to p and q respectively and

$$
B \;:=\; \begin{pmatrix} 0 & 1 & 1 \\ 1 & 1 & 0 \\ (-1 \cdot p) & q & (p + q) \end{pmatrix}.
$$

– $A_{(p \cdot q)} := (A_p, A_q, B)$ where A_p and A_q are the matrices corresponding to p and q respectively and

$$B := \begin{pmatrix} 0 & 1 & 1 \\ 1 & q & 0 \\ (-1 \cdot p) & 0 & (p \cdot q) \end{pmatrix}.$$

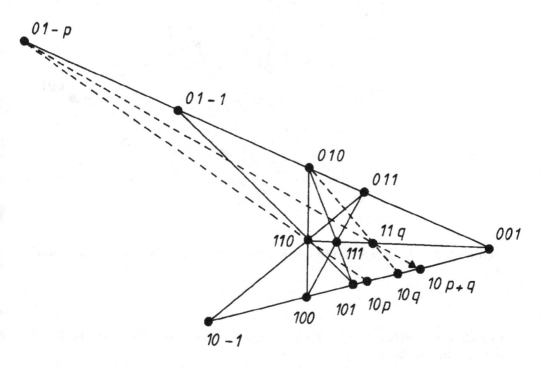

Figure 2-1. Projective addition: geometric interpretation of the matrix $A_{(p+q)}$.

A geometric interpretation of this construction method is given in the Figures 2-1 and 2-2. Let A_p be any such matrix with entries from $L_n \cup \{0\}$. Consider all entries of A_p as elements of the field $K'(x_1, \ldots, x_n)$, and compute the corresponding *simple* rank 3 matroid M_p. More precisely: two non-zero columns of M_p are identified if they are scalar multiples, and redundant columns are deleted to obtain a reduced $3 \times N$ matrix $\widehat{A}_p$. W.l.o.g. we can assume that the first 9 columns equal the frame matrix A. M_p is the rank 3 linear matroid on $E = \{1, 2, \ldots, N\}$ of the matrix $\widehat{A}_p$ with respect to the field $K'(x_1, \ldots, x_n)$.

It follows from the construction of A_p that the matroid M_p is projectively unique with respect to any field extension of the prime field K'. That implies

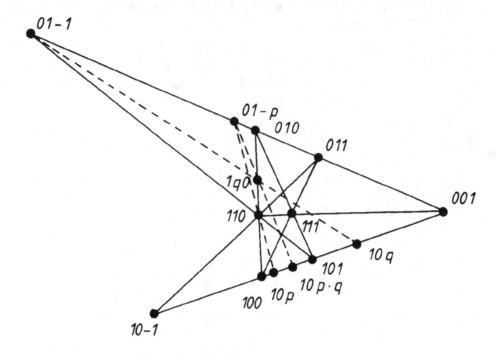

Figure 2-2. Projective multiplication: geometric interpretation of the matrix $A_{(p \cdot q)}$.

that after a suitable projective transformation, every K-coordinate matrix for M_p contains the submatrix

$$\begin{pmatrix} 1 & \ldots & 1 & 1 & \ldots & 1 & \ldots & 1 \\ 0 & \ldots & 0 & 0 & \ldots & 0 & \ldots & 0 \\ 0 & \ldots & a_1 & a_2 & \ldots & a_n & \ldots & p(a_1, \ldots, a_n) \end{pmatrix}$$

where $a_1, a_2, \ldots a_n \in K$. Now we can force the first column $(1,0,0)^T$ and last column $(1,0,p(a_1, \ldots, a_n))^T$ to be equal by identifying the two corresponding points in the matroid M_p on $E = \{1, 2, \ldots, N\}$. More precisely : let $\widehat{M_p}$ denote rank 3 matroid on $\hat{E} = \{1, 2, \ldots, N-1\}$ where $b \subset \hat{E}$ is a basis of $\widehat{M_p}$ if and only if b is a basis in M_p or $(b \setminus 1) \cup N$ is a basis in M_p. (In a very few special cases it can happen that $\widehat{M_p}$ is not even a matroid. Still it is a "collection of bases" which can have matroids as "minors".)

The new matroid $\widehat{M}_p$ is still projectively unique, and hence after a suitable projective transformation every K-coordinate matrix for $\widehat{M}_p$ contains the submatrix

$$\begin{pmatrix} 1 & \ldots & 1 & 1 & \ldots & 1 \\ 0 & \ldots & 0 & 0 & \ldots & 0 \\ 0 & \ldots & a_1 & a_2 & \ldots & a_n \end{pmatrix}$$

where $a_1, a_2, \ldots a_n \in K$ necessarily satisfy the equation $p(a_1, \ldots, a_n) = 0$. Hence $\widehat{M}_p$ is already a matroid which is K-realizable only if $p = 0$ has roots in K.

In order to turn this "only if" into an "if and only if", we have to consider weak images of $\widehat{M}_p$. For, $p(a_1, \ldots, a_n) = 0$ might have a solution $a_1, a_2, \ldots, a_n \in K$, which, when written into matrix form as above, implies additional dependencies and thus does not give coordinates for the matroid M_p. For example, a_i might equal 1 or $a_i = a_j$ for some $i \neq j$. In this case it could be possible that $p = 0$ has such a (degenerate) solution but $\widehat{M}_p$ is still not K-coordinatizable.

We have to be careful in picking the necessary weak images, because if "too degenerate" weak images are allowed, then the nine-point coordinate frame might become degenerate as well. For the projective addition and multiplication to make sense, we need to ensure that the 9-point minor M_A corresponding to the frame matrix A is still a minor. Let $\mathcal{M}_p$ denote the set of weak images of M_p for which the corresponding weak map induces an isomorphism on the frame minor M_A. All objects being finite, the set $\mathcal{M}_p$ is computable.

At this point our argument is completed for the case of one polynomial $\mathcal{F} = \{p\}$ because it follows from the construction that there exists a member in $\mathcal{M}_p$ which is K-coordinatizable if and only if the equation $p = 0$ has roots in K. It is clear, however, that all procedures can be generalized to a finite set $\mathcal{F} = \{p_1, \ldots, p_n\}$ of polynomial expressions by assigning to this set a coordinate matrix $A_{\mathcal{F}}$ by concatenation of the above recursively defined coordinated matrices A_{p_i}. This proves Theorem 2.2. $\qquad\Box$

2.2. On the algorithmic Steinitz problem

Here we investigate the existence of algorithms for the Steinitz problem and the coordinatizability problem of oriented matroids. Due to their nature these problems make sense only over an *ordered* field K. It is the main goal of this section to show that these two problems are algorithmically equivalent. The reduction of the Steinitz problem to oriented matroids is very easy from a theoretical point of view because it amounts to simply enumerating the finite set of all matroid polytopes with given fixed face lattice. Several computational aspects of this approach have been discussed in [32]; in that paper the polytopality problem for a large class of spheres has been solved via the reduction to oriented matroids. At the end of this section we describe an algorithm to compute the matroid polytopes for a given sphere which improves the procedure in [24, Section 3].

The theoretically more difficult step is the reduction of oriented matroid co-ordinatizability to the polytopality of spheres. Our solution to this problem is based on a technique due to J. Lawrence that assigns to every oriented matroid χ an oriented matroid $\Lambda(\chi)$ whose face lattice is polytopal over K if and only if χ is coordinatizable over K [12, Theorem 2.2]. To begin with, note the following easy observation.

Remark 2.3. *For any ordered field K, we have the implications $(1) \Rightarrow (2) \Rightarrow (3)$ among the following statements.*

(1) There exists an algorithm for deciding any finite set of polynomials $f_1, \ldots, f_r, g_1, \ldots, g_s \in \mathbf{Z}[x_1, \ldots, x_n]$, $r, s, n \in \mathbf{N}$, whether there exist $a_1, \ldots, a_n \in K$ such that $f_i(a_1, \ldots, a_n) = 0$ for $i = 1, \ldots, r$ and $g_j(a_1, \ldots, a_n) > 0$ for $j = 1, \ldots, s$.

(2) There exists an algorithm for deciding an arbitrary oriented matroid χ whether χ is coordinatizable over K.

(3) There exists an algorithm for deciding an arbitrary matroid M whether M is coordinatizable over K.

Proof. The implication $(1) \Rightarrow (2)$ follows immediately from the definitions in Section 1.3. An oriented matroid χ is K-coordinatizable if and only if the system of equations and inequalities (with integer coefficients) defined by its chirotope has a solution over K.

To prove $(2) \Rightarrow (3)$ we have to use a given decision procedure for K-coordinatizability of oriented matroids in constructing such a procedure for matroids. Given any matroid M, let $\mathcal{O}(M)$ denote the set of oriented matroids with underlying matroid M. All sets being finite objects, $\mathcal{O}(M)$ is computable. M is K-coordinatizable if and only if there exists a K-coordinatizable oriented matroid $\chi \in \mathcal{O}(M)$. $\square$

Let χ be a rank r oriented matroid on $E = \{e_1, \ldots, e_n\}$, and let $E = \{e'_1, \ldots, e'_n\}$ be a disjoint copy of E. The *Lawrence extension* of χ is the unique rank $2n - r$ oriented matroid $\Lambda(\chi)$ on $E \cup E'$ such that

(i) $\Lambda(\chi)/E' = \chi^*$, the dual of χ, and

(ii) $\{e, e'\}$ is a positive cocircuit of $\Lambda(\chi)$ for all $e \in E$.

Clearly, the mapping $\chi \mapsto \Lambda(\chi)$ is computable. For different but equivalent definitions of the Lawrence extension, see [12, Section 2] or [24, Section 6]. It also directly follows from any definition of the Lawrence extension that for a given ordered field K, χ is K-coordinatizable if and only if $\Lambda(\chi)$ is K-coordinatizable.

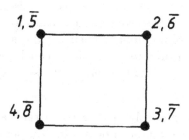

Figure 2-3. The dual to the Lawrence extension $\Lambda(\chi)$ of a quadrangle. The co-facets of the prism over a tetrahedron are the positive circuits in this diagram, e.g. $\{1,5\}$ or $\{1,3,6,8\}$.

For example, if χ is the rank 3 oriented matroid of a quadrangle then $\Lambda(\chi)$ is the oriented matroid associated with a 4-polytope, namely the prism over a tetrahedron, see Figure 2-3.

The following lemma is a direct consequence of the fact that, by condition (ii), the vertex set $E \cup E'$ of the matroid polytope $\Lambda(\chi)$ can be partitioned into 2-element co-facets, compare [11], [12, Lemma 2.1], [126, Section 3].

Lemma 2.4. *Given any oriented matroid χ, the Lawrence extension $\Lambda(\chi)$ is rigid, that is, it is uniquely determined by its face lattice.*

Now we are prepared to prove the main result of this section.

Theorem 2.5. *For any ordered field K the following statements are equivalent*
(1) There exists an algorithm for deciding an arbitrary oriented matroid χ whether χ is coordinatizable over K.
(2) There exists an algorithm for deciding an arbitrary (finite) lattice $\mathcal{L}$ whether $\mathcal{L}$ is the face lattice of a convex polytope P with vertex coordinates in K.

Proof. To prove $(1) \Rightarrow (2)$ assume there exists a decision procedure for K-coordinatizability of oriented matroids. For any lattice $\mathcal{L}$ we can define

$$\mathcal{O}_{\mathcal{L}} \quad := \quad \{ \chi \mid \chi \text{ is a matroid polytope with face lattice } \mathcal{L} \}.$$

Clearly, $\mathcal{L}$ is K-polytopal if and only if there exists a K-coordinatizable $\chi \in \mathcal{O}_{\mathcal{L}}$. Since there are only finitely many matroid polytopes with fixed rank and number of vertices, the finite set $\mathcal{O}_{\mathcal{L}}$ can be obtained by enumerating all such matroid polytopes and testing whether their face lattices are isomorphic to $\mathcal{L}$.

Conversely, assume we are given an algorithm for deciding an arbitrary finite lattice whether it is K-polytopal. Then it follows from Lemma 2.4 that the K-coordinatizability of an oriented matroid χ can be checked by applying this algorithm to the face lattice of its Lawrence extension $\Lambda(\chi)$. □

Enumerating all matroid polytopes of given rank and number of vertices is certainly a very inefficient method to determine the set $\mathcal{O}_{\mathcal{L}}$ for a given lattice $\mathcal{L}$. An efficient algorithm for this task has been described in [24, Section 3]. Here we would like to suggest a different approach.

Let us first assume for simplicity that $\mathcal{L}$ is the face lattice of a pure $(r-2)$-dimensional *simplicial* complex with n vertices. The set $\mathcal{F}_{\mathcal{L}}$ of maximal simplices (= facets) of $\mathcal{L}$ can be considered as a subset of $\Lambda(n, r-1)$. If $\mathcal{L}$ is polytopal, then $\mathcal{L}$ is necessarily a triangulated sphere and hence orientable. Thus sorting the facets appropriately gives us an *intrinsic orientation* $i_{\mathcal{L}} : \mathcal{F}_{\mathcal{L}} \to \{-1, +1\}$.

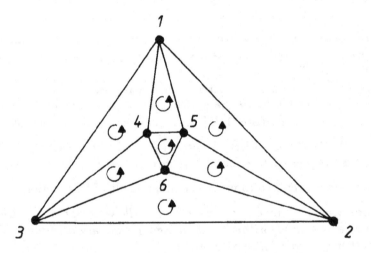

Figure 2-4. The intrinsic orientation $i_{\mathcal{L}}$ of the face lattice of an octahedron.

For example, if $\mathcal{L}$ is the face lattice of an octahedron, then

$$\mathcal{F}_{\mathcal{L}} = \{[123], [134], [145], [125], [236], [346], [456], [256]\}.$$

The intrinsic orientation $i_{\mathcal{L}}$ of this simplicial complex is given by

$$[123] \mapsto 1, \quad [134] \mapsto 1, \quad [145] \mapsto 1, \quad [125] \mapsto -1,$$
$$[236] \mapsto -1, \quad [346] \mapsto -1, \quad [456] \mapsto -1, \quad [256] \mapsto 1,$$

see Figure 2-4.

This intrinsic $(r-1)$-dimensional orientation $i_{\mathcal{L}}$ together with the "convex embedding condition" defines a unique extrinsic r-dimensional orientation for $\mathcal{L}$, that is, a unique partial oriented matroid $\chi_{\mathcal{L}}$.

$$\chi_{\mathcal{L}} : \Lambda(n,r) \rightarrow \{-1,+1,*\}$$
$$\lambda \mapsto \begin{cases} (-1)^j \cdot i_{\mathcal{L}}(\lambda_1,..,\lambda_{j-1},\lambda_{j+1},..,\lambda_d) & \text{if } \lambda \setminus \{\lambda_j\} \text{ is facet of } \mathcal{L} \\ * & \text{else} \end{cases}$$

The fact that $i_{\mathcal{L}}$ is the intrinsic orientation of an orientable simplicial complex guarantees that $\chi_{\mathcal{L}}$ is well-defined: for a "simplex" λ which is adjacent to two facets (or more), both facets yield the same result in the formula defining $\chi_{\mathcal{L}}$.

In the case of the above octahedron, for example, the following partial oriented matroid $\chi_{\mathcal{L}}$ is derived from the above intrinsic orientation $i_{\mathcal{O}}$.

[1234] +	[1235] +	[1236] +	[1245] +	[1246] *
[1256] −	[1345] +	[1346] +	[1356] *	[1456] +
[2345] *	[2346] +	[2356] +	[2456] +	[3456] +

It directly follows from our construction that $\widetilde{\chi_{\mathcal{L}}} = \mathcal{O}_{\mathcal{L}}$ where $\mathcal{O}_{\mathcal{L}}$ is as in the proof of Theorem 2.5 and $\widetilde{\chi_{\mathcal{L}}}$ stands for the set of all oriented matroids above $\chi_{\mathcal{L}}$, see Section 1.3.

Let us now briefly discuss the general case, i.e. the complex associated with the lattice $\mathcal{L}$ is not assumed to be simplicial any longer. We then consider the *order complex* $\widehat{\mathcal{L}}$ of $\mathcal{L}$, that is the simplicial complex of all maximal chains of $\mathcal{L}$ ordered by inclusion. Geometrically speaking, $\widehat{\mathcal{L}}$ is the *first barycentric subdivision* [87] of the complex $\mathcal{L}$. For this simplicial complex a partial oriented matroid $\chi_{\widehat{\mathcal{L}}}$ is defined as above. In deleting all (new) subdividing points in $\chi_{\widehat{\mathcal{L}}}$, we obtain a partial oriented matroid $\Pi_{\mathcal{L}} : \Lambda(n,r) \rightarrow \{-1,+1,*\}$.

Again $\Pi_{\mathcal{L}}$ can be considered to be an extrinsic orientation of the complex associated with $\mathcal{L}$. In the non-simplicial case we need to ensure, however, that every non-simplicial facet is contained in a hyperplane. If we alter $\Pi_{\mathcal{L}}$ by imposing this condition, then we will obtain all (partial) oriented matroids with face lattice $\mathcal{L}$. We define

$$\chi_{\mathcal{L}} : \Lambda(n,r) \rightarrow \{-1,0,+1,*\}$$
$$\lambda \mapsto \begin{cases} 0 & \text{if } \lambda \subset F \text{ for some facet } F \text{ of } \mathcal{L}, \\ \Pi_{\mathcal{L}}(\lambda) & \text{else.} \end{cases}$$

It is easy to see that if $\mathcal{L}$ is the face lattice of some matroid polytope χ, then $\widehat{\mathcal{L}}$ is the face lattice of an extension $\widehat{\chi}$ of χ. In fact $\widehat{\chi}$ can be obtained from χ by successive *principal extensions* [125, Theorem 6.19]. It follows from this argument that also in the general case $\widetilde{\chi_{\mathcal{L}}} = \mathcal{O}_{\mathcal{L}}$, i.e. the oriented matroids with face lattice $\mathcal{L}$ are precisely the oriented matroids above $\chi_{\mathcal{L}}$. $\chi_{\mathcal{L}}$ is K-coordinatizable if and only if the set $\widetilde{\chi_{\mathcal{L}}}$ contains a K-coordinatizable oriented matroid. This proves in particular the following statement, which will be of importance in Chapter IV in describing an algebraic criterion for the polytopality of lattices.

Proposition 2.6. *Given any matroid sphere $\mathcal{S}$, there exists a partial oriented matroid $\chi_{\mathcal{S}}$ such that $\mathcal{S}$ is K-polytopal if and only if $\chi_{\mathcal{S}}$ is K-coordinatizable.*

2.3. Real realizability versus rational realizability

In the first two sections of this chapter we have dealt with an arbitrary coordinate field (or ordered field) K. In this full generality we were only able to prove statements of the form *"If problem (A) can be solved algorithmically over K, so can problem (B)"*, and so far we did not say anything about whether there really exist algorithms for the problems in question. In this section we discuss the important case of real closed fields, compare Section 4.1, and we prove that the decidability of our synthetic problems over the rational numbers is equivalent to the unsolved rational version of Hilbert's 10th problem.

Many synthetic problems and realizability results related to finite fields are of mainly combinatorial nature. This holds in particular for the beautiful and deep results of Tutte and others on $\mathbf{GF_p}$-coordinatizability of matroids that we mentioned earlier in Section 1.3. Note also that the solvability of algebraic equations over finite fields is a finite problem and hence trivially decidable.

Having in mind the applications from Section 1.1, it can be stated that, in general, computational synthetic geometry deals with realizations over subfields of the real numbers $\mathbf{R}$, or at least, in characteristic 0. Realizations over the complex numbers or over fields of rational functions can be of interest from an algebraic point of view or because they provide additional insight into real realizations, and so, we aim to prove our results for arbitrary fields whenever possible. Yet, from a strictly logical point of view we cannot compute even with real numbers, but we have to restrict ourselves to the real algebraic numbers $\mathbf{A}$.

The most important reason for our focus on real algebraic numbers is that they are in a sense universal for all ordered fields. It is a consequence of Tarski's famous theorem on the completeness of the theory of real closed fields, see [88], [154], that a system of equations and inequalities with integer coefficients has a solution over some ordered field if and only if it has a solution within $\mathbf{A}$. The consequences of this result for convex polytopes, line arrangements and geometric complexes have been discussed by Grünbaum [60, Chapter 5], [61, Theorem 2.27] and Lindström [106]: if these structures are realizable over some ordered field, then they are realizable over $\mathbf{A}$.

The other field of main interest is the "smallest" ordered field, the rationals $\mathbf{Q}$. Realizations over $\mathbf{Q}$ are motivated by the "diophantine problems" that we mentioned in the introduction.

Here we prove that for these two fields the realizability of matroids and oriented matroids and the Steinitz problem are equivalent to solving a single polynomial equation.

Theorem 2.7. *Let K denote either the rationals $\mathbf{Q}$ or its real closure, the real algebraic numbers $\mathbf{A}$. Then the following statements are equivalent.*

(1) *There exists an algorithm for deciding an arbitrary polynomial $f \in \mathbf{Z}[x_1, \ldots, x_n]$, $n \in \mathbf{N}$, whether f has zeros in K^n.*

(2) *There exists an algorithm to decide for an arbitrary matroid M whether M is coordinatizable over K.*

(3) *There exists an algorithm to decide for an arbitrary oriented matroid χ whether χ is coordinatizable over K.*

(4) *There exists an algorithm to decide for an arbitrary finite lattice $\mathcal{L}$ whether $\mathcal{L}$ is isomorphic to the face lattice of a convex polytope with vertex coordinates in K.*

Let us remark that for $K = \mathbf{A}$, the real algebraic numbers, statement (1) is known to be true by the above mentioned result of Tarski [154]. Tarski's *Quantifier Elimination* method, which is only of theoretical interest, can today be replaced by much more efficient methods, in particular Collins' *Cylindrical Algebraic Decomposition* algorithm [43] and the decision procedure of Grigor'ev and Vorobjev [75]. For related complexity results see also Ben-Or et.al. [13] and Canny [42].

The situation changes entirely if we focus our attention on the field $K = \mathbf{Q}$ of rational numbers. Matiyasevic's negative solution [112] to Hilbert's 10th problem in 1971, stating that "there exists no algorithm for deciding whether a system of diophantine equations has a solution among the rational *integers*," cannot be applied to prove the corresponding statement for rational *numbers*, and as B. Mazur points out in a survey article [113], this problem is still open, see also Klee & Wagon [94]. So the case $K = \mathbf{Q}$ in Theorem 2.7 implies the remarkable result that the rational version of Hilbert's 10th problem is equivalent to the rational Steinitz problem.

Problem 2.8. (Hilbert's 10th problem, rational version) *Does an algorithm exists for deciding an arbitrary polynomial $f \in \mathbf{Q}[x_1, \ldots, x_n]$, $n \in \mathbf{N}$, whether f has zeros in $\mathbf{Q}^n$.*

In order to prove Theorem 2.7 we need the two following lemmas.

Lemma 2.9. *Let K be an ordered field such that for some fixed positive integer $\nu = \nu(K)$ the condition*

$$\forall x \in K \quad (\ x \geq 0 \quad \Longleftrightarrow \quad \exists x_1, \ldots, x_\nu \in K : x = \sum_{k=1}^{\nu} x_k^2 \ ) \qquad (P)$$

is satisfied. If there exists an algorithm for deciding an arbitrary polynomial $f \in \mathbf{Z}[x_1, \ldots, x_n]$, $n \in \mathbf{N}$, whether f has zeros in K^n, then there also exists an algorithm for deciding any finite set of polynomials $f_1, \ldots, f_r, g_1, \ldots, g_s \in \mathbf{Z}[x_1, \ldots, x_n]$, $r, s, n \in \mathbf{N}$, whether there exist $a_1, \ldots, a_n \in K$ such that $f_i(a_1, \ldots, a_n) = 0$ for $i = 1, \ldots, r$ and $g_j(a_1, \ldots, a_n) > 0$ for $j = 1, \ldots, s$.

Proof. First replace the inequalites $g_j > 0$ by the equations $h_j = 0$ as follows. For each $j = 1, \ldots, s$ introduce ν new variables y_{jk}, $k = 1, \ldots, \nu$, with ν as in condition (P). Define

$$h_j(x_1, \ldots, x_n, y_{j1}, \ldots, y_{j\nu}) \quad := \quad g_j(x_1, \ldots, x_n) \cdot \sum_{k=1}^{\nu} y_{jk}^2 \quad - \quad 1.$$

According to condition (P), $g_j > 0$ has a solution over K if and only if $h_j = 0$ has a solution over K. Finally, define

$$f(x_1, .., x_n, y_{11}, .., y_{s\nu}) \quad := \quad \sum_{i=1}^{r} f_i(x_1, .., x_n)^2 \quad + \quad \sum_{j=1}^{s} h_j(x_1, .., x_n, y_{j1}, .., y_{j\nu})^2.$$

Since every ordered field is formally real, see Section 4.1, a sum of squares in K vanishes if and only if each summand vanishes. This shows that the original system of equations and inequalities has a solution K^n if and only if the equation $f = 0$ has a solution in $K^{n+s \cdot \nu}$. $\qquad \square$

Observe that for the real numbers $\mathbf{R}$ and the real algebraic numbers $\mathbf{A}$ condition (P) holds with $\nu = 1$.

Lemma 2.10. *The field $\mathbf{Q}$ of rationals satisfies the condition (P) and hence the statement of Lemma 2.9.*

Proof. By *Lagrange's four-square theorem* [67, Section 20.5], every positive integer can be written as a sum of four squares. From this it easily follows that every positive rational number is the sum of at most $\nu = 4$ squares of rational numbers. For, given $p, q \in \mathbf{N}$ with $p \cdot q = \sum_{i=1}^{4} r_i^2$ in $\mathbf{N}$, we have

$$\frac{p}{q} \quad = \quad \frac{p \cdot q}{q^2} \quad = \quad \sum_{i=1}^{4} \left(\frac{r_i}{q}\right)^2.$$

Proof of Theorem 2.7. We only need to assemble the parts we have proved so far. By Theorem 2.2 we have $(2) \iff (1)$, by the second part of Remark 2.3 we have $(3) \Rightarrow (2)$, and by Theorem 2.5 we have $(4) \iff (3)$. The implication $(1) \Rightarrow (3)$ follows from Lemma 2.9, Lemma 2.10 and the first part of Remark 2.3.

$\square$

Chapter III

COMBINATORIAL AND ALGEBRAIC METHODS

In this chapter we discuss combinatorial and algebraic methods for the realizability problems of matroids and oriented matroids. For the sake of simplicity we emphasize the rank 3 case. The first two sections deal with (unoriented) matroids: a coordinatization method based on Grassmann algebra is developed and applied to various examples, in particular, we classify the algebraic varieties corresponding to all types of 10_3-configurations. We also study inequality reductions for oriented matroids, and we describe in detail the construction of coordinates and a non-realizability proof for a realizable and a non-realizable example respectively. For a history of this reduction technique of Bokowski see Chapter VIII. The results for the examples in Section 3.3 have been obtained by Sturmfels in joint work with D.Ljubić & J.-P.Roudneff [108] and J.Bokowski & J.Richter [28a] respectively.

3.1. A basis-free coordinatization algorithm for a class of matroids

In order to coordinatize a rank 3 matroid M on $E = \{1, 2, \ldots, n\}$ over a field K, we need to establish an $(n \times 3)$-matrix $A = (a_{ij})$ with entries from K such that for all $\lambda \in \binom{E}{3}$

$$\det \begin{pmatrix} a_{\lambda_1 1} & a_{\lambda_1 2} & a_{\lambda_1 3} \\ a_{\lambda_2 1} & a_{\lambda_2 2} & a_{\lambda_2 3} \\ a_{\lambda_3 1} & a_{\lambda_3 2} & a_{\lambda_3 3} \end{pmatrix} = 0$$

if and only if λ is dependent in M, which amounts to solving a system of equations in $3n$ variables. If a matrix A describes a coordinatization of a matroid M, so does every matrix B which is projectively equivalent to A, i.e. $B = D \cdot A \cdot T$ where D is a non-singular diagonal $n \times n$-matrix and T a non-singular $d \times d$-matrix. This argument can be used to simplify the equations and to reduce the number of variables to $2(n - 4)$, see e.g. Fenton [63].

We have seen in Chapter II that arbitrary "bad" polynomial systems can be obtained in this way. In any case, at that point one is left with a purely algebraic problem, and theoretically the system can be solved by any algorithm that solves algebraic equations over K, e.g. over $\mathbf{C}$ by the Gröbner bases method [40], [99], [150] and over $\mathbf{R}$ by Collins' method. It is very likely, however, that in practice this only works for very small n, while for larger n these algorithms can solve only small systems corresponding to matroids which are trivially coordinatizable. It is the great disadvantage of the straightforward algebraization of a geometry problem that one is forced to forget the underlying geometric structure.

In this section we suggest an invariant-theoretic (hence geometric) coordinatization technique for a large class of matroids. Although the basic idea works for higher-dimensional problems as well, we restrict ourselves to the rank 3 case.

The method is based on the *Grassmann algebra* (in the literature also refered to as *double algebra, Cayley algebra, ...*) [48], [55], [152].

In the following we will use the abbreviations $U \vee V$ for the sum and $U \wedge V$ for the intersection of two linear subspaces U, V of K^d. Given two linearly independent vectors $a, b \in K^d$, we identify both vectors with their mutual linear spans, and we write $a \vee b$ for the plane spanned by both of them.

Given three vectors x, y and z in K^3, K any field, the determinant $\det(x, y, z)$ is abbreviated as before by the *bracket* $[xyz]$. Here $[xyz]$ will be thought of as a multilinear operator in 3 vector-valued variables. All geometric properties of linear dependencies, hence the entire information stored in a matroid, can be expressed in terms of brackets. Moreover, it follows from the First Fundamental Theorem of Invariant Theory (Theorem 1.7) that all computations necessary for coordinatizing a rank 3 matroid can only be carried out using only bracket expressions, compare [43, Section 9], W. Whiteley, Dissertation 1971, Harward University. It is one of the aims of this chapter to show that this is not only a purely theoretical observation but that in many instances brackets can be used for very efficient calculations.

To begin with an example, let us decide the coordinatizability of the rank 3 matroid M_8 with $n = 8$ points given in Figure 3-1.

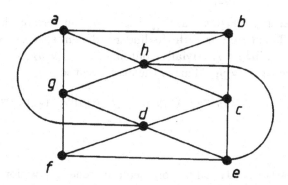

Figure 3-1. The "complex" matroid M_8.

Can we find a set $E := \{a, b, c, d, e, f, g, h\} \subset K^3$ of 8 vectors such that

$$[abd] = [bce] = [cdf] = [deg] = [efh] = [fga] = [ghb] = [hac] = 0 \quad (1)$$

and such that $[xyz] \neq 0$ for all other $\{x, y, z\} \in \binom{E}{3}$? Let us proceed inductively and assume we had a coordinatization of $M \setminus a$, that is, vectors $b, c, \ldots, h \in K^3$ such that

$$[bce] = [cdf] = [deg] = [efh] = [ghb] = 0. \quad (2)$$

In general, however, this configuration cannot be extended by another vector a to a coordinatization of M. In fact, this is possible if and only if the planes $b \vee d$, $f \vee g$ and $h \vee c$ are in a special position: they have to intersect in a line.

Proposition 3.1. Let $x_1, y_1, x_2, y_2, x_3, y_3$ be vectors in general position in K^3, i.e. any three of them are linearly independent. Then the planes $x_1 \vee y_1$, $x_2 \vee y_2$ and $x_3 \vee y_3$ have a non-zero vector in common if and only if

$$[x_1 y_1 x_2][y_2 x_3 y_3] + [y_1 x_1 y_2][x_2 x_3 y_3] = 0.$$

Let us first prove the following lemma.

Lemma 3.2. If x_1, y_1, x_2, y_2 are vectors in general position in K^3, then the planes $x_1 \vee y_1$ and $x_2 \vee y_2$ intersect in the line spanned by the non-zero vector $v := [x_1 y_1 x_2] \cdot y_2 + [y_1 x_1 y_2] \cdot x_2$.

Proof. The expressions $[x_1 y_1 x_2]$ and $[y_1 x_1 y_2]$ are non-zero by the assumption of general position. This together with the linear independence of x_2 and y_2 implies that v is non-zero. Since v is trivially contained in $x_2 \vee y_2$, we need only show that v is contained in $x_1 \vee y_1$. This is the case because

$$[x_1 y_1 v] = [x_1 y_1 x_2][x_1 y_1 y_2] + [x_1 y_1 y_2][y_1 x_1 x_2] = 0.$$

$\square$

Proof of Proposition 3.1. All vectors being in general position, the lines $(x_1 \vee y_1) \wedge (x_2 \vee y_2)$ and $(x_3 \vee y_3) \wedge (x_2 \vee y_2)$ are well-defined and they are spanned by $v_1 := [x_1 y_1 x_2] \cdot y_2 + [y_1 x_1 y_2] \cdot x_2$ and $v_3 := [x_3 y_3 x_2] \cdot y_2 + [y_3 x_3 y_2] \cdot x_2$ respectively.

The planes $x_1 \vee y_1$, $x_2 \vee y_2$ and $x_3 \vee y_3$ have a non-zero vector in common if and only if v_1 and v_3 are linearly dependent. This, however, is the case if and only

$$[x_1 y_1 x_2][x_3 y_3 y_2] + [y_1 x_1 y_2][x_3 y_3 x_2] = 0.$$

$\square$

These two formulas suffice to solve the system (1). By an initial application of Proposition 3.1 we can reduce (1) to (2), plus one additional bracket identity. Thereafter, we can use Lemma 3.2 to express b as a bracket linear combination of g and h. Repeated application of Lemma 3.2 leads to a situation where no vector is contained in more than one vanishing bracket, and then we have to start introducing parameters. Finally, we are left with four variable vectors in general position for which we can, by projective equivalence, assume any coordinatization. The parameters chosen along the way yield an implicit description of the space of all weak coordinatizations of M_8 as the zero set of a single polynomial with integer coefficients, from which the coordinatizability properties of M_8 can be read off.

Here is a complete transcript of this reduction for M_8.

Vanishing brackets = non-bases of the matroid :
[abd], [bce], [cdf], [deg], [efh], [fga], [ghb], [hac]

Solving for a by Propostion 3.1 :
[bdf][ghc] + [dbg][fhc] = 0

Replace b by [ceg] h + [ech] g :
[([ceg]h + [ech]g) d f][ghc] + [d ([ceg] h + [ech] g) g][fhc]=

[ceg][hdf][ghc] + [ech][gdf][ghc] +
[ceg][dhg][fhc] + [ech][dgg][fhc]   =

[ceg][hdf][ghc] + [ech][gdf][ghc] + [ceg][dhg][fhc]

Replace e by [d g f] h + [g d h] f :
[c ([dgf]h + [gdh]f) g][h d f][g h c] +
[([dgf]h + [gdh]f) c h][g d f][g h c] +
[c ([dgf]h + [gdh]f) g][d h g][f h c]   =

[d g f][c h g][h d f][g h c] + [g d h][c f g][h d f][g h c] +
[d g f][h c h][g d f][g h c] + [g d h][f c h][g d f][g h c] +
[d g f][c h g][d h g][f h c] + [g d h][c f g][d h g][f h c]   =

[d g f][c h g][h d f][g h c] + [g d h][c f g][h d f][g h c] +
[g d h][f c h][g d f][g h c] + [d g f][c h g][d h g][f h c] +
[g d h][c f g][d h g][f h c]

Replace f by ν c + μ d , where ν and μ are parameters
[d g (νc + μd)][c h g][h d (νc + μd)][g h c] +
[g d h][c (νc + μd) g][h d (νc + μd)][g h c] +
[g d h][(νc + μd) c h][g d (νc + μd)][g h c] +

$[d \ g \ (\nu c \ + \ \mu d)] \, [c \ h \ g] \, [d \ h \ g] \, [(\nu c \ + \ \mu d) \ h \ c] \ +$
$[g \ d \ h] \, [c \ (\nu c \ + \ \mu d) \ g] \, [d \ h \ g] \, [(\nu c \ + \ \mu d) \ h \ c] \quad =$

$\nu \ \nu \ [dgc] \, [chg] \, [hdc] \, [ghc] \ + \ \nu \ \mu \ [dgc] \, [chg] \, [hdd] \, [ghc] \ +$
$\mu \ \nu \ [dgd] \, [chg] \, [hdc] \, [ghc] \ + \ \mu \ \mu \ [dgd] \, [chg] \, [hdd] \, [ghc] \ +$
$\nu \ \nu \ [gdh] \, [ccg] \, [hdc] \, [ghc] \ + \ \nu \ \mu \ [gdh] \, [ccg] \, [hdd] \, [ghc] \ +$
$\mu \ \nu \ [gdh] \, [cdg] \, [hdc] \, [ghc] \ + \ \mu \ \mu \ [gdh] \, [cdg] \, [hdd] \, [ghc] \ +$
$\nu \ \nu \ [gdh] \, [cch] \, [gdc] \, [ghc] \ + \ \nu \ \mu \ [gdh] \, [cch] \, [gdd] \, [ghc] \ +$
$\mu \ \nu \ [gdh] \, [dch] \, [gdc] \, [ghc] \ + \ \mu \ \mu \ [gdh] \, [dch] \, [gdd] \, [ghc] \ +$
$\nu \ \nu \ [dgc] \, [chg] \, [dhg] \, [chc] \ + \ \nu \ \mu \ [dgc] \, [chg] \, [dhg] \, [dhc] \ +$
$\mu \ \nu \ [dgd] \, [chg] \, [dhg] \, [chc] \ + \ \mu \ \mu \ [dgd] \, [chg] \, [dhg] \, [dhc] \ +$
$\nu \ \nu \ [gdh] \, [ccg] \, [dhg] \, [chc] \ + \ \nu \ \mu \ [gdh] \, [ccg] \, [dhg] \, [dhc] \ +$
$\mu \ \nu \ [gdh] \, [cdg] \, [dhg] \, [chc] \ + \ \mu \ \mu \ [gdh] \, [cdg] \, [dhg] \, [dhc] \quad =$

$\nu \ \nu \ [dgc] \, [chg] \, [hdc] \, [ghc] \ + \ \mu \ \nu \ [gdh] \, [cdg] \, [hdc] \, [ghc] \ +$
$\mu \ \nu \ [gdh] \, [dch] \, [gdc] \, [ghc] \ + \ \nu \ \mu \ [dgc] \, [chg] \, [dhg] \, [dhc] \ +$
$\mu \ \mu \ [gdh] \, [cdg] \, [dhg] \, [dhc]$

Finally, by projective equivalence, we can assume w.l.o.g. that
$[cdg] \ = \ 1 \quad [dgh] \ = \ 1 \quad [ghc] \ = \ 1 \quad [hcd] \ = \ 1$
and the coordinatizability is equivalent to the identity
$\nu \ \nu \ + \ \mu \ \nu \ - \ \mu \ \nu \ + \ \nu \ \mu \ + \ \mu \ \mu \ = \ \nu^2 \ + \ \nu\mu \ + \ \mu^2 \ = \ 0 \ .$

This identity has non-trivial solutions within the complex numbers but not within the real numbers, and hence M_8 is **C**-coordinatizable but not **R**-coordinatizable. In fact, M_8 is coordinatizable over a field K if and only if the polynomial $x^2 + x + 1$ factors in $K[x]$.

Let us summarize the reduction method which led to this polynomial. It is clear that this technique will not work for matroids with "too many" dependencies, without making "too many" precise at this point. In the next section we shall see that the class of matroids for which the algorithm works is quite large.

Algorithm 3.3.

Input : Rank 3 matroid M with ≥ 4 points.

Output : Polynomial $p \in \mathbf{Z}[\nu_1, \ldots, \nu_r]$ such that the weak coordinatizations of M (modulo projective equivalence and with respect to a fixed basis) correspond to the points on the hypersurface $p = 0$ in K^r.

1. $p := 0$.
2. If M has only four points e_1, e_2, e_3 and e_4, then replace the bracket $[e_i e_j e_k]$ in p by $\text{sign}(ijk)$, i.e. by either $+1$ or -1. Print p and STOP.
3. Pick $e \in E$ such that the number $n(e)$ of non-bases of M containing e is minimal. Write $\{e, x_i, y_i\}$, $i = 1, \ldots, n(e)$ for the non-bases of M containing e.

4.1. If $p = 0$ and $n(e) \leq 2$, then set $M := M \setminus e$ and go to 2.

4.2. If $p = 0$ and $n(e) = 3$, then set $p := [x_1 y_1 x_2][y_2 x_3 y_3] + [y_1 x_1 y_2][x_2 x_3 y_3]$, set $M := M \setminus e$ and go to 2.

4.3. If $p \neq 0$ and $n(e) = 0$, replace each occurrence of e in p by $\lambda_e x + \mu_e y + \nu_e z$, $\{x, y, z\} \subset E$ any basis, set $M := M \setminus e$ and go to 2.

4.4. If $p \neq 0$ and $n(e) = 1$, then replace each occurrence of e in p by $\mu_e x_1 + \nu_e y_1$, set $M := M \setminus e$ and go to 2.

4.5. If $p \neq 0$ and $n(e) = 2$, then replace each occurrence of e in p by $[x_1 y_1 x_2] \cdot y_2 + [y_1 x_1 y_2] \cdot x_2$, set $M := M \setminus e$ and go to 2.

4.6. Otherwise, STOP; the algorithm does not apply to M.

3.2. Applications to n_3-configurations and other examples

We call a rank 3 matroid M with n points an n_k-*configuration*, $k \geq 3$, if every line of M contains at most k points and every point is contained in precisely k k-point lines [47], [78], [85], [142], [152]. We can think of such an M as being represented by a bipartite k-regular graph on the union $P \cup L$ of the set P of points and the set L of k-point lines of M, and conversely, every bipartite k-regular graph (satisfying a weak extra condition) yields an n_k-configuration.

"*It might be mentioned that there was a time when the study of (projective) configurations was considered the most important branch of all geometry*" [69, Chapter 3]. This quotation from the famous book on Geometry and the Intuition by Hilbert and Cohn-Vossen seems nearly unbelievable today, in a time when projective geometry is a highly endangered species. From the algorithmic point of view, however, things look different, and the constructive aspects of old-fashioned projective geometry (as embodied in configurations, matroids, polyhedra and related structures) can hardly be overestimated in their significance for computational geometry.

The case $k = 3$ is of particular interest in the study of n_k-configurations; one reason being that the two fundamental theorems of projective geometry, the theorems of *Pappus* and *Desargues*, can be expressed in terms of n_3-configurations. It should be added that among all matroids with non-trivial coordinatization properties the n_3-configurations are the "least degenerate" ones, and so it is very natural to study n_3-configurations under the aspect of coordinatizability.

Remark 3.4. *Algorithm 3.3 works for all n_3-configurations.*

In fact, as a byproduct we obtain straightforward automated proofs for the theorems of Pappus and Desargues. When applying Algorithm 3.3 to the corresponding matroids, the proof consists of the reduction to "$0 = 0$" of the polynomial condition in the coordinatization parameters ($\mu_e, \nu_e, \ldots$ as introduced throughout the procedure).

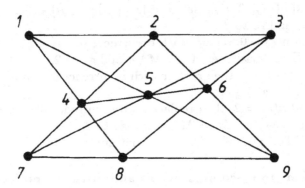

Figure 3-2. The Pappus configuration $(9_3)_1$.

Example 3.5. *The Pappus Configuration* $(9_3)_1$.
We display the transcript of Algorithm 3.3 as a proof of Pappus' theorem. The matroid of the Pappus-9_3-configuration is given by its nonbases :　[1 2 3],
[1 4 8],[1 5 9],[2 4 7],[2 6 9],[3 5 7],[3 6 8],[4 5 6],[7 8 9]

Applying 4.2. to e = 1 yields :
[2 3 4][8 5 9] + [3 2 8][4 5 9] = p

Applying 4.5. to e = 2 and simplification yields :
[9 3 4][4 7 6][8 5 9] + [6 3 4][7 4 9][8 5 9] +
[3 9 8][4 7 6][4 5 9] + [3 6 8][7 4 9][4 5 9] = p

Applying 4.5. to e = 3 and simplification yields :
[9 8 4][5 7 6][4 7 6][8 5 9] + [9 6 4][7 5 8][4 7 6][8 5 9] +
[6 8 4][5 7 6][7 4 9][8 5 9] + [6 9 8][7 5 8][4 7 6][4 5 9]= p

Applying 4.4. to e = 4, simplifying and factoring out ν_4 [5 7 6] yields :
ν_4 [9 8 5][5 7 6][8 5 9] + μ_4 [9 8 6][5 7 6][8 5 9] +
ν_4 [9 6 5][7 5 8][8 5 9] + ν_4 [6 8 5][7 5 9][8 5 9] +
μ_4 [6 8 5][7 6 9][8 5 9] + μ_4 [6 9 8][7 5 8][6 5 9]　=　p

Applying 4.4. to e = 7 and simplification yields p = 0 as desired :
ν_4 ν_7 [9 8 5][5 8 6][8 5 9] + ν_4 μ_7 [9 8 5][5 9 6][8 5 9] +
μ_4 ν_7 [9 8 6][5 8 6][8 5 9] + μ_4 μ_7 [9 8 6][5 9 6][8 5 9] +
ν_4 μ_7 [9 6 5][9 5 8][8 5 9] + ν_4 ν_7 [6 8 5][8 5 9][8 5 9] +
μ_4 ν_7 [6 8 5][8 6 9][8 5 9] + μ_4 μ_7 [6 9 8][9 5 8][6 5 9]　=　0

A similar but slightly longer computation works also for the Desargues 10_3- configuration, given by its nonbases [1 2 5], [1 3 6], [1 4 8], [2 3 7], [2 4 9], [3 4 0], [5 6 7], [5 8 9], [6 8 0], [7 9 0] .

In Section 3.1 we have seen that the unique 8_3-configuration (it can be shown that there is only one up to isomorphism) is coordinatizable over the complex numbers but not over the real numbers. Here are two more interesting examples which illustrate the importance of the underlying coordinate field.

Example 3.6. *The Fano plane 7_3.*

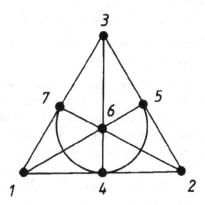

Figure 3-3. The Fano plane 7_3.

Nonbases : [124], [235], [346], [457], [561], [672], [713]

Applying 4.2. to $e =$ 1 yields : $p =$[245][673] + [426][573]

Applying 4.5. to $e =$ 2 and simplification yields : $p =$
[745][356][673] + [645][537][673] + [476][356][573]

Applying 4.5. to $e =$ 4 and simplification yields : $p =$
[567][637][356][357] + [567][637][356][357]

Since the brackets [5 6 7], [6 3 7], [3 5 6], [3 5 7] are supposed to be non-zero, we obtain as a necessary coordinatizability condition for the Fano-plane 7_3 that $1+1 = 2 = 0$, i.e. the characteristic of the coordinate field has to be 2. In fact, 7_3 can be coordinatized over $\mathbf{GF_2}$: it is isomorphic to the matroid associated with $(\mathbf{GF_2})^3 \setminus \{0\}$.

Example 3.7. *An irrational matroid, see Figure 1-2.*
This is a smallest example of a matroid which can be coordinatized with real numbers but *not with rationals.* Its non-bases are [1 2 3], [1 2 4], [1 5 9], [1 6 7], [2 5 8], [2 6 9], [3 6 8], [3 7 9], [4 5 6], [4 7 8]

In applying Algorithm 3.3 with the sequence $e = 3, e = 1, e = 2, e = 7, e = 4$, we obtain the polynomial

$$p = \mu_4^2 - \mu_4 \cdot \nu_4 - \nu_4^2$$

which has non-trivial real roots but no rational roots. Hence the matroid has the desired property of being irrational, (cf. [33]).

Finally, we prove the following result, compare Conjecture 1.3.

Theorem 3.8. *Given $n \leq 10$, every n_3-configuration which is real realizable is also realizable with rational coordinates.*

Proof. For $n \leq 9$, the result follows from the discussion in [69, Chapter 3]. Up to isomorphism, there are ten combinatorial types of 10_3-configurations which have been classified by H. Schröter [135] (see also [78]). The configuration $(10_3)_1$ is the Desargues-configuration which is clearly rationally realizable. The fourth configuration in Schröter's list is the configuration $(10_3)_4$ which is known to be not realizable over any (commutative) field, [47], [77], [103]. With Algorithm 3.3 we obtained an automated non-realizability proof for $(10_3)_4$ which will be omitted here because of its length (ca. 3 pages).

For all other eight configurations we found integer coordinates with the aid of Algorithm 3.3. The derivations will be listed in Figures 3-4 to 3-11. These tables are to be read as follows: the combinatorial description of the configurations, which is the input for Algorithm 3.3, is given in line (1). The output, that is the polynomial p, is given in (2). Now *all* realizations (over any field) are obtained by choosing a root of the polynomial p in (2) and performing the operations described in the list (3), which is obtained automatically (in reverse order) as a transcript of the application of Algorithm 3.3.

At this point we know that (2) and (3) together yield a parameterization of the space of all (weak) coordinatizations (with respect to the specified projective basis). The main problem now was to find an integer solution to the equation $p = 0$ such that the corresponding coordinatization is proper, i.e. no additional dependencies occur. For the configurations $(10_3)_2$ and $(10_3)_3$ that was easy because it is possible to write one parameter as rational function of the other four. This also implies that for these two configurations the rational solutions are dense in the real solutions, see Conjecture 1.4. The other six equations have been solved by hand calculations, supported by a trial-and-error computer program. The integer solutions are given in (4), and the corresponding integer coordinate matrix (5) is obtained by plugging the values from (4) into the scheme (3). □

In the meantime Sturmfels and White [152] have applied a modified version of Algorithm 3.3 to all 11_3- and 12_3-configurations in the lists of R. Daublebsky [51], [52]. Such an investigation was suggested by B. Grünbaum. They obtained the following result.

Theorem 3.9. *All 11_3-configurations and all 12_3-configurations (in the classification of R. Daublebsky) are realizable with rational coordinates.*

Here is another problem of Grünbaum in the same spirit. In [81] he describes a real coordinatization of a 21_4-configuration, and he asks

Problem 3.10. *Is there a real realizable n_4-configuration for some $n \leq 20$?*

$$[123], [150], [189], [257], [268], [369], [370], [456], [478], [490] \tag{1}$$

$$-\alpha_5^2 - \alpha_5\alpha_6\beta_5 + \alpha_5\beta_5\gamma_6 + \alpha_6\beta_5^2 + \alpha_6\beta_5^2\gamma_6 + \beta_5^2\beta_6^2 + \beta_5^2\beta_6\gamma_6 \tag{2}$$

$$x_7 := (1,0,0)^T, \quad x_8 := (0,1,0)^T, \quad x_9 := (0,0,1)^T, \quad x_0 := (1,-1,1)^T$$

$$x_6 := \alpha_6 x_7 + \beta_6 x_8 + \gamma_6 x_9$$

$$x_5 := \alpha_5 x_4 + \beta_5 x_6 \tag{3}$$

$$x_4 := [789]x_0 + [870]x_9$$

$$x_3 := [697]x_0 + [960]x_7$$

$$x_2 := [576]x_8 + [758]x_6$$

$$x_1 := [508]x_9 + [059]x_8$$

$$\alpha_5 := 3, \quad \beta_5 := 1, \quad \alpha_6 := -4, \quad \beta_6 := 1, \quad \gamma_6 := 3 \tag{4}$$

$$\begin{pmatrix} 0 & 12 & 4 & 1 & -1 & -4 & 1 & 0 & 0 & 1 \\ -3 & 6 & -1 & -1 & -2 & 1 & 0 & 1 & 0 & -1 \\ 4 & -9 & 1 & 0 & 3 & 3 & 0 & 0 & 1 & 1 \end{pmatrix} \tag{5}$$

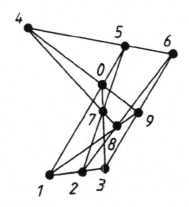

Figure 3-4. Rational coordinatization of the configuration $(10_3)_2$.

$$[139], [146], [158], [237], [240], [269], [350], [457], [680], [789] \tag{1}$$

$$(\beta_4 + \gamma_4)(-\alpha_4^2\alpha_6 + \alpha_4^2\beta_6 + 2\alpha_4\beta_4\beta_6 - \alpha_6\beta_4\gamma_4 + \beta_4^2\beta_6) \tag{2}$$

$$x_5 := (1,0,0)^T, \quad x_8 := (0,1,0)^T, \quad x_9 := (0,0,1)^T, \quad x_0 := (1,-1,1)^T$$

$$x_4 := \alpha_4 x_5 + \beta_4 x_8 + \gamma_4 x_9$$

$$x_6 := \alpha_6 x_8 + \beta_6 x_0$$

$$x_7 := [458]x_9 + [549]x_8$$

$$x_2 := [406]x_9 + [049]x_6$$

$$x_3 := [275]x_0 + [720]x_5$$

$$x_1 := [465]x_8 + [648]x_5$$

$$\alpha_4 := 2, \quad \beta_4 := 4, \quad \gamma_4 := 71, \quad \alpha_6 := 1, \quad \beta_6 := 8 \tag{4}$$

$$\begin{pmatrix} -552 & 48 & -3600 & 2 & 1 & 8 & 0 & 0 & 0 & 1 \\ 529 & -42 & 3450 & 4 & 0 & -7 & 4 & 1 & 0 & -1 \\ 0 & 117 & -3450 & 71 & 0 & 8 & 71 & 0 & 1 & 1 \end{pmatrix} \tag{5}$$

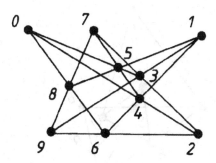

Figure 3-5. Rational coordinatization of the configuration $(10_3)_3$.

$$[124], [138], [179], [237], [259], [350], [456], [480], [678], [690] \tag{1}$$

$$-\alpha_3^2\alpha_5\beta_5^2\gamma_5 + \alpha_3^2\beta_5^2\gamma_5^2 + \alpha_3\alpha_5^2\beta_3\gamma_5 + 3\alpha_3\alpha_5\beta_3\beta_5\gamma_5 - \alpha_3\alpha_5\beta_3\gamma_5^2 + \alpha_3\beta_3\beta_5^2\gamma_5$$

$$-2\alpha_3\beta_3\beta_5\gamma_5^2 + \alpha_5^2\beta_3^2 + 2\alpha_5\beta_3^2\beta_5 - \alpha_5\beta_3^2\gamma_5 + \beta_3^2\beta_5^2 - \beta_3^2\beta_5\gamma_5 \tag{2}$$

$$x_7 := (1,0,0)^T, \quad x_8 := (0,1,0)^T, \quad x_9 := (0,0,1)^T, \quad x_0 := (1,-1,1)^T$$

$$x_5 := \alpha_5 x_7 + \beta_5 x_8 + \gamma_5 x_9$$

$$x_3 := \alpha_3 x_5 + \beta_3 x_0$$

$$x_6 := [789]x_0 + [870]x_9$$

$$x_4 := [568]x_0 + [650]x_8$$

$$x_2 := [375]x_9 + [739]x_5$$

$$x_1 := [387]x_9 + [839]x_7$$

$$\alpha_3 := 2, \quad \beta_3 := 3, \quad \alpha_5 := 23, \quad \beta_5 := 12, \quad \gamma_5 := 3 \tag{4}$$

$$\begin{pmatrix} -49 & 483 & 49 & 3 & 23 & 1 & 1 & 0 & 0 & 1 \\ 0 & 252 & 21 & 32 & 12 & -1 & 0 & 1 & 0 & -1 \\ -9 & 108 & 9 & 3 & 3 & 0 & 0 & 0 & 1 & 1 \end{pmatrix} \tag{5}$$

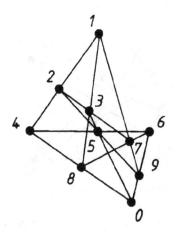

Figure 3-6. Rational coordinatization of the configuration $(10_3)_5$.

$$[123], [149], [150], [248], [257], [360], [379], [456], [678], [890] \qquad (1)$$

$$(\alpha_4\alpha_8 + \alpha_4\beta_8 - \gamma_4\beta_8)(\beta_4\alpha_8 + \beta_4\beta_8 + \gamma_4\beta_8 + \gamma_4\alpha_8)\beta_4 +$$

$$(\beta_4\alpha_8 + \beta_4\beta_8 + \gamma_4\beta_8)\gamma_4\alpha_8\beta_4 - (\beta_4\alpha_8\gamma_4\alpha_8 + \beta_4\beta_8\gamma_4\alpha_8 + \gamma_4\beta_8\gamma_4\alpha_8)(\beta_4 + \alpha_4) \quad (2)$$

$$x_5 := (1,0,0)^T, \quad x_7 := (0,1,0)^T, \quad x_9 := (0,0,1)^T, \quad x_0 := (1,-1,1)^T$$

$$x_4 := \alpha_4 x_5 + \beta_4 x_7 + \gamma_4 x_9$$

$$x_8 := \alpha_8 x_9 + \beta_8 x_0$$

$$x_6 := [457]x_8 + [548]x_7$$

$$x_3 := [607]x_9 + [069]x_7$$

$$x_2 := [485]x_7 + [847]x_5$$

$$x_1 := [234]x_9 + [329]x_4$$

$$\alpha_8 := 1, \quad \beta_8 := 2, \quad \alpha_4 := 10, \quad \beta_4 := 6, \quad \gamma_4 := 9 \qquad (4)$$

$$\begin{pmatrix} -10 & 12 & 0 & 10 & 1 & 18 & 0 & 2 & 0 & 1 \\ -6 & 36 & 36 & 6 & 0 & 18 & 1 & -2 & 0 & -1 \\ 6 & 0 & 9 & 9 & 0 & 27 & 0 & 3 & 1 & 1 \end{pmatrix} \qquad (5)$$

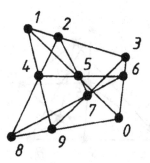

Figure 3-7. Rational coordinatization of the configuration $(10_3)_6$.

$$[123], [148], [159], [247], [260], [356], [379], [450], [678], [890] \tag{1}$$

$$\alpha_4^2 \alpha_5^2 \beta_5 \gamma_5 + \alpha_4^2 \alpha_5^2 \gamma_5^2 + \alpha_4^2 \alpha_5 \beta_5^2 \gamma_5 + \alpha_4^2 \alpha_5 \beta_5 \gamma_5^2$$

$$+ \alpha_4 \alpha_5^2 \beta_4 \gamma_5 + \alpha_4 \alpha_5 \beta_4 \beta_5 \gamma_5 - \alpha_5 \beta_4^2 \beta_5 - \beta_4^2 \beta_5^2 \tag{2}$$

$$x_6 := (1,0,0)^T, \quad x_7 := (0,1,0)^T, \quad x_9 := (0,0,1)^T, \quad x_0 := (1,-1,1)^T$$

$$x_5 := \alpha_5 x_6 + \beta_5 x_7 + \gamma_5 x_9$$

$$x_4 := \alpha_4 x_5 + \beta_4 x_0$$

$$x_8 := [679]x_0 + [760]x_9$$

$$x_3 := [567]x_9 + [659]x_7$$

$$x_2 := [476]x_0 + [740]x_6$$

$$x_1 := [485]x_9 + [849]x_5$$

$$\alpha_4 := 1, \quad \beta_4 := 3, \quad \alpha_5 := 1, \quad \beta_5 := 88, \quad \gamma_5 := 8 \tag{4}$$

$$\begin{pmatrix} 89 & -4 & 0 & 4 & 1 & 1 & 0 & 1 & 0 & 1 \\ 7832 & 11 & 88 & 85 & 88 & 0 & 1 & -1 & 0 & -1 \\ 979 & -11 & 8 & 11 & 8 & 0 & 0 & 0 & 1 & 1 \end{pmatrix} \tag{5}$$

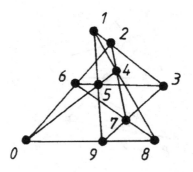

Figure 3-8. Rational coordinatization of the configuration $(10_3)_7$.

$$[123], [146], [150], [248], [279], [345], [369], [578], [670], [890] \tag{1}$$

$$-\alpha_4^2\alpha_8\beta_4\beta_8 - \alpha_4^2\beta_4\beta_8^2 - \alpha_4^2\beta_8^2\gamma_4 - \alpha_4\alpha_8^2\beta_4^2 - 2\alpha_4\alpha_8\beta_4^2\beta_8 + \alpha_4\alpha_8\beta_8\gamma_4^2$$
$$-\alpha_4\beta_4^2\beta_8^2 + \alpha_4\beta_8^2\gamma_4^2 + \alpha_8\beta_4^2\beta_8\gamma_4 + \alpha_8\beta_4\beta_8\gamma_4^2 + \beta_4^2\beta_8^2\gamma_4 + \beta_4\beta_8^2\gamma_4^2 \tag{2}$$

$$x_5 := (1,0,0)^T, \quad x_6 := (0,1,0)^T, \quad x_9 := (0,0,1)^T, \quad x_0 := (1,-1,1)^T$$

$$x_4 := \alpha_4 x_5 + \beta_4 x_6 + \gamma_4 x_9$$

$$x_8 := \alpha_8 x_9 + \beta_8 x_0$$

$$x_7 := [586]x_0 + [850]x_6$$

$$x_3 := [456]x_9 + [549]x_6$$

$$x_2 := [487]x_9 + [849]x_7$$

$$x_1 := [465]x_0 + [640]x_5$$

$$\alpha_8 := 15, \quad \beta_8 := 3, \quad \alpha_4 := 3, \quad \beta_4 := 2, \quad \gamma_4 := 4 \tag{4}$$

$$\begin{pmatrix} -3 & -270 & 0 & 3 & 1 & 0 & -18 & 3 & 0 & 1 \\ 4 & 45 & 2 & 2 & 0 & 1 & 3 & -3 & 0 & -1 \\ -4 & -990 & 4 & 4 & 0 & 0 & -18 & 18 & 1 & 1 \end{pmatrix} \tag{5}$$

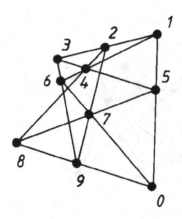

Figure 3-9. Rational coordinatization of the configuration $(10_3)_8$.

$$[138], [159], [160], [234], [268], [279], [357], [456], [470], [890] \tag{1}$$

$$-\alpha_5^2\alpha_8^2\beta_5 - 2\alpha_5^2\alpha_8\beta_5\beta_8 - \alpha_5^2\alpha_8\beta_8\gamma_5 - \alpha_5^2\beta_5\beta_8^2 - \alpha_5^2\beta_8^2\gamma_5$$

$$+\alpha_5\alpha_8^2\beta_5\gamma_5 - \alpha_5\alpha_8\beta_5^2\beta_8 + \alpha_5\alpha_8\beta_5\beta_8\gamma_5 - \alpha_5\beta_5^2\beta_8^2 - \alpha_5\beta_5\beta_8^2\gamma_5 - \alpha_8\beta_5\beta_8\gamma_5^2 \tag{2}$$

$$x_6 := (1,0,0)^T, \quad x_7 := (0,1,0)^T, \quad x_9 := (0,0,1)^T, \quad x_0 := (1,-1,1)^T$$

$$x_5 := \alpha_5 x_6 + \beta_5 x_7 + \gamma_5 x_9$$

$$x_8 := \alpha_8 x_9 + \beta_8 x_0$$

$$x_4 := [567]x_0 + [650]x_7$$

$$x_2 := [687]x_9 + [869]x_7$$

$$x_3 := [245]x_7 + [427]x_5$$

$$x_1 := [596]x_0 + [950]x_6$$

$$\alpha_8 := 19, \quad \beta_8 := 2, \quad \alpha_5 := 2, \quad \beta_5 := 3, \quad \gamma_5 := 16 \tag{4}$$

$$\begin{pmatrix} -2 & 0 & 672 & 16 & 2 & 1 & 0 & 2 & 0 & 1 \\ -3 & 2 & -322 & 3 & 3 & 0 & 1 & -2 & 0 & -1 \\ 3 & -21 & 5376 & 16 & 16 & 0 & 0 & 21 & 1 & 1 \end{pmatrix} \tag{5}$$

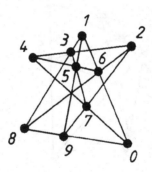

Figure 3-10. Rational coordinatization of the configuration $(10_3)_9$.

$$[123], [157], [169], [240], [258], [345], [379], [468], [670], [890] \tag{1}$$

$$\alpha_4^2\alpha_8\beta_8\gamma_4 - \alpha_4\alpha_8^2\beta_4^2 - 2\alpha_4\alpha_8^2\beta_4\gamma_4 - 2\alpha_4\alpha_8\beta_4^2\beta_8 - 2\alpha_4\alpha_8\beta_4\beta_8\gamma_4 - 2\alpha_4\alpha_8\beta_8\gamma_4^2$$

$$-\alpha_4\beta_4^2\beta_8^2 - \alpha_4\beta_4\beta_8^2\gamma_4 - \alpha_8^2\beta_4^2\gamma_4 - \alpha_8\beta_4\beta_8\gamma_4^2 + \beta_4^2\beta_8^2\gamma_4 + \beta_4\beta_8^2\gamma_4^2 \tag{2}$$

$$x_5 := (1,0,0)^T, \quad x_7 := (0,1,0)^T, \quad x_9 := (0,0,1)^T, \quad x_0 := (1,-1,1)^T$$

$$x_4 := \alpha_4 x_5 + \beta_4 x_7 + \gamma_4 x_9$$

$$x_8 := \alpha_8 x_9 + \beta_8 x_0$$

$$x_6 := [487]x_0 + [840]x_7$$

$$x_3 := [457]x_9 + [549]x_7$$

$$x_2 := [405]x_8 + [048]x_5$$

$$x_1 := [576]x_9 + [759]x_6$$

$$\alpha_8 := 5, \quad \beta_8 := 16, \quad \alpha_4 := 4, \quad \beta_4 := 16, \quad \gamma_4 := 9 \tag{4}$$

$$\begin{pmatrix} -60 & 500 & 0 & 4 & 1 & 60 & 0 & 16 & 0 & 1 \\ 160 & -400 & 16 & 16 & 0 & -160 & 1 & -16 & 0 & -1 \\ 0 & 525 & 9 & 9 & 0 & 60 & 0 & 21 & 1 & 1 \end{pmatrix} \tag{5}$$

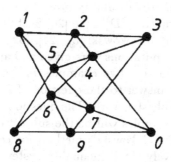

Figure 3-11. Rational coordinatization of the configuration $(10_3)_1 0$.

3.3. Inequality reduction and coordinatization of oriented matroids

We have seen in Chapter II that deciding the realizability of arbitrary (oriented) matroids is as difficult as solving arbitrary polynomial equations and inequalities with integer coefficients, and theoretically it would be sufficient to refer to any decision procedure for real closed fields, e.g. Collins' Cylindrical Algebraic Decomposition [43]. Computer experiments of B.Kutzler at the University of Linz [private communication], however, show that in most "real" geometric applications the corresponding inequality system is still too large for the versions of Collins' method presently implemented. On the other hand, the very specific structure of most systems arising from applications gives some hope that substantial improvements can be expected from a collaboration of Computer Algebra and Computational Synthetic Geometry.

In this section we discuss some techniques for the realizability problem of uniform oriented matroids, the basic ideas of which are due to J.Bokowski. His inequality reduction method, underlying Algorithm 3.10, has been further developed and programmed in joint work, see [28a]. For a systematic algorithmic approach to the coordinatizability of oriented matroids, see also [31]. Observe that many techniques in this section as well as in [31] generalize to non-uniform oriented matroids, too.

In order to show the application of these methods to two non-trivial examples, we shall discuss in detail one realizable and one non-realizable rank 3 oriented matroid. These two oriented matroids are respectively due to J.P.-Roudneff [108] and J.Richter [28a]. The realizability proof for Roudneff's oriented matroid χ_{12} was given by Sturmfels and D. Ljubić and is contained in the article [108].

We present an algebraic non-realizability proof by constructing a *final polynomial* for Richter's oriented matroid D_3^{10}, see [20, Sect.2]. The final polynomial method has been introduced by J.Bokowski [5], [24], [32]. In these papers, however, only the results of the computations were given and applied to the geometric problem in question. By discussing two examples in detail in this section, we attempt to give a satisfactory answer to the question of *how to find a final polynomial* for a given non-realizable structure. The question of the existence of final polynomials and many more examples will be the subject of Chapter IV.

To begin with, we describe Bokowski's method for constructing a small reduced system, that is a relatively small inequality system that still carryies the entire information for a given uniform rank d oriented matroid χ with n points. More precisely, $\mathcal{R} \subset \Lambda(n,d)$ is called a *reduced system* for χ if χ is uniquely determined by its restriction to $\mathcal{R}$, i.e. $\chi'|_{\mathcal{R}} = \chi|_{\mathcal{R}}$ implies $\chi' = \chi$ for every oriented matroid χ'.

We first consider brackets $[\lambda]$ which are necessarily contained in every reduced system of χ. These brackets have been studied under the names *mutations* in [131] and *invertible bases* in [14]. We use the abbreviation $\{\sigma|\tau\} :=$

$\{\sigma_1 \ldots \sigma_{d-2} \mid \tau_1 \ldots \tau_4\}$ for the *three term Grassmann-Plücker syzygy*

$$
\begin{aligned}
& [\sigma_1 \ldots \sigma_{d-2}\tau_1\tau_2] \cdot [\sigma_1 \ldots \sigma_{d-2}\tau_3\tau_4] \\
- \ & [\sigma_1 \ldots \sigma_{d-2}\tau_1\tau_3] \cdot [\sigma_1 \ldots \sigma_{d-2}\tau_2\tau_4] \\
+ \ & [\sigma_1 \ldots \sigma_{d-2}\tau_1\tau_4] \cdot [\sigma_1 \ldots \sigma_{d-2}\tau_2\tau_3]
\end{aligned}
$$

where $\tau \in \Lambda(n,4), \sigma \in \Lambda(n, d-2)$. Recall that these syzygies are sufficient to define oriented matroids in the uniform case [31]. A syzygy $\{\sigma \mid \tau\}$ is said to *determine* a tuple $[\sigma\tau_i\tau_j]$ in χ if $\chi(\sigma, \tau_i, \tau_j)$ is uniquely determined by the values of χ for the other 5 tuples occuring in $\{\sigma \mid \tau\}$ and the oriented matroid condition for that syzygy (Definition 1.12). The *closure* $< \mathcal{D} >$ of a subset $\mathcal{D} \subset \Lambda(n,d)$ is the union of $\mathcal{D}$ with the set of brackets $[\sigma, \tau_i, \tau_j] \in \Lambda(n,d)$ which are determined in χ by some syzygy $\{\sigma \mid \tau\}$ whose other five brackets are already determined by $\mathcal{D}$.

The idea is now to find a "small" subset $\mathcal{D} \subset \Lambda(n,d)$ with $< \mathcal{D} > = \Lambda(n,d)$. $\lambda \in \Lambda(n,d)$ is a *mutation* if it is not determined by any three-term syzygy. We write $\mathrm{Mut}(\chi)$ for the set of mutations of χ. By the results of [131], the mutations of a uniform rank d oriented matroid χ are in one-to-one correspondence to the simplicial regions of the associated $(d-1)$-arrangement of pseudohyperplanes [64].

While there are oriented matroids χ for which $\mathrm{Mut}(\chi)$ is already a reduced system, this cannot be expected in general. An example of this phenomenon is given in Figure 5-1.

All later computations are most conveniently carried out with respect to a basis $\beta \in \Lambda(n,d)$, and so it is reasonable to add the set

$$
V_\beta \ := \ \{ \ [\beta_1, \ldots, \beta_{i-1}, k, \beta_{i+1}, \ldots, \beta_d] \ \mid \ k \notin \beta, i \in \{1, \ldots, d\} \ \}
$$

of "variables" (with respect to β) to the set of mutations. Observe that here we are free to choose among $\binom{n}{d}$ possible bases.

Remark 3.11. *Given any* $\lambda \in \Lambda(n,d)$, *the expansion of* $[\lambda]$ *with respect to a basis* β *reads*

$$
[\lambda] \cdot [\beta]^{d-1} \ = \ \det \begin{pmatrix}
[\lambda_1\beta_2 \ldots \beta_d] & [\beta_1\lambda_1 \ldots \beta_d] & \ldots & [\beta_1\beta_2 \ldots \lambda_1] \\
[\lambda_2\beta_2 \ldots \beta_d] & [\beta_1\lambda_2 \ldots \beta_d] & \ldots & [\beta_1\beta_2 \ldots \lambda_2] \\
\vdots & \vdots & \ddots & \vdots \\
[\lambda_d\beta_2 \ldots \beta_d] & [\beta_1\lambda_d \ldots \beta_d] & \ldots & [\beta_1\beta_2 \ldots \lambda_d]
\end{pmatrix}
$$

The above $d \times d$-determinant reduces to a $k \times k$-determinant where $k = |\lambda \backslash \beta|$. To choose a suitable basis β we introduce a certain monotone "weight" function $w : \mathbf{N} \to \mathbf{N}$. Let $\beta \in \Lambda$ such that the expression $v(\beta) := \sum_{\lambda \in \mathrm{Mut}(\chi)} w(|\lambda \backslash \beta|)$ is minimized. In practice the following two weight functions turned out to be most useful.

– *Counting the number of "too big" determinants.*

$$w(k) \;=\; \begin{cases} 0 & \text{if } k \le m \\ 1 & \text{if } k > m \end{cases}$$

where m is fixed.
– *Counting the total number of occurrences of variables in determinants.*
$w(k) := k^2$.

Now the following "filling up" algorithm determines a small reduced system.

Algorithm 3.12.
Input : *A uniform oriented matroid* $\chi : \Lambda(n,d) \to \{-1,+1\}$, $\beta \in \Lambda(n,d)$.
Output : *Small reduced system* $\mathcal{R}$ *for* χ.

1. Let $\mathcal{R} := \mathrm{Mut}(\chi) \cup V_\beta$, $\mathcal{M} := \emptyset$.
2. Determine $\mathcal{D} := <\mathcal{R}>$.
3. If $\mathcal{D} = \Lambda(n,d)$,
   3.1 Then GO TO 5.
   3.2. Else : Pick $\mu \in \Lambda(n,d) \setminus \mathcal{D}$ such that $|\mu \setminus \beta|$ is minimal.
4. Let $\mathcal{R} := \mathcal{R} \cup \{\mu\}$, $\mathcal{M} := \mathcal{M} \cup \{\mu\}$. Go to 2.
5. If $\mathcal{M} := \emptyset$,
   5.1 Then STOP, $\mathcal{D}$ is a reduced system.
   5.2 Else: Pick $\lambda \in \mathcal{M}$, $\mathcal{M} := \mathcal{M} \setminus \{\lambda\}$
   5.3 If $<\mathcal{D} \setminus \{\lambda\}> = \Lambda(n,d)$, then $\mathcal{R} = \mathcal{R} \setminus \{\lambda\}$.
   5.4 GO TO 5.

In the first example we consider the pseudo-line arrangement $\mathcal{A}'_{12}$ depicted in Figure 3-12. In [79] B.Grünbaum posed the problem of whether there exists an arrangement of (straight) lines with the property that no two of its triangles are adjacent. The pseudo-line arrangement $\mathcal{A}'_{12}$ which is due to J.-P.Roudneff [108] has exactly this property, and so Grünbaum's problem would be solved if $\mathcal{A}'_{12}$ were stretchable.

J.-P. Roudneff constructed the arrangement $\mathcal{A}'_{12}$ from the unique simple arrangement of 5 pseudo-lines by recursively applying a procedure which, at each step, reduces the number of adjacent pairs of triangles. The arrangement $\mathcal{A}'_{12}$ has 67 faces, more precisely, 16 triangles, 39 quadrangles and 12 pentagons. The 16 triangles can be read off from Table 3-1 where they correspond to the underlined triples. Observe also that all proper subarrangements and all arrangements obtainable from $\mathcal{A}'_{12}$ by "switching" the orientation of one triangle do have adjacent triangles which shows how difficult it is to find such an arrangement.

Let us now outline the combinatorial and algebraic reduction steps which led to the stretchability decision for $\mathcal{A}'_{12}$. Using the correspondence of Folkman and Lawrence [64], we assign to $\mathcal{A}'_{12}$ a uniform rank 3 oriented matroid χ_{12} :

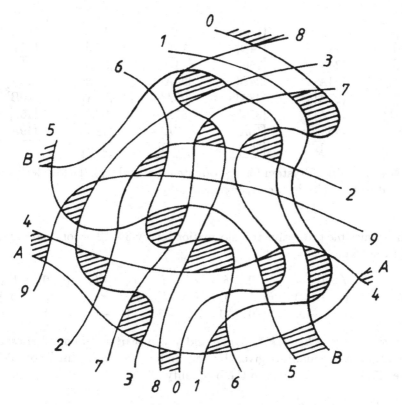

Figure 3-12. The (stretchable) arrangement $\mathcal{A}'_{12}$ of 12 pseudo-lines without adjacent triangles.

$\{1, 2, \ldots, 9, 0, A, B\}^3 \rightarrow \{-1, +1\}$ such that χ_{12} is realizable if and only if $\mathcal{A}'_{12}$ is stretchable.

In using Algorithm 3.12, we determine a reduced system $\mathcal{R}$ for χ_{12} which is listed in Table 3-1. The system $\mathcal{R}$ contains 46 of the $\binom{12}{3} = 220$ bases. Recall that, by the results of [131], the triangles of $\mathcal{A}'_{12}$ are necessarily contained in every reduced system $\mathcal{R}$ of χ_{12}.

[124]+	[125]+	[126]+	[127]+	[128]+	[12A]+
[13A]+	[14A]−	[14B]+	[16A]−	[170]+	[234]−
[23A]+	[245]−	[246]−	[247]−	[248]−	[249]−
[240]+	[24A]+	[24B]+	[25A]+	[269]+	[26A]+
[278]−	[27A]+	[28A]+	[29A]+	[20A]−	[20B]−
[2AB]+	[34A]+	[359]−	[37A]−	[38B]+	[450]−
[45A]−	[468]−	[46A]−	[47A]−	[48A]−	[49A]+
[40A]+	[4AB]−	[567]+	[5AB]+	[80A]−	

Table 3-1. A reduced system for the oriented matroid χ_{12}. The underlined bases correspond to triangles in the arrangement $\mathcal{A}'_{12}$.

After an admissible projective transformation, every 3×12−coordinate matrix for χ_{12} is of the form

$$
A \;=\; \begin{pmatrix} -1 & -1 & a & 0 & b & c & d & e & -f & -g & 0 & -h \\ -1 & 0 & i & 1 & j & k & l & m & n & -o & 0 & -p \\ 1 & 0 & 1 & 0 & -1 & -1 & -1 & -1 & -1 & 1 & 1 & 1 \end{pmatrix}
$$

where $a, b, c, d, e, f, g, h, i, j, k, l, m, n, o$ and p are *positive* real numbers satisfying the determinant inequalities given by the underlined triples in Table 3-1. For example, [567]+ corresponds to the inequality

$$
\det \begin{pmatrix} b & c & d \\ j & k & l \\ -1 & -1 & -1 \end{pmatrix} \;=\; (c-b)(k-l) - (d-c)(j-k) \;>\; 0.
$$

When 12 trivial inequalities and 16 inequalities which express the positivity of the variables are deleted, we are left with a system of 19 inequalities in 16 positive variables as listed in Table 3-2. Let us remark that as a consequence of the application of Algorithm 3.12, all expressions of the form (...) in the last four inequalities are necessarily positive.

1	<	j	[125]+
1	<	k	[126]+
1	<	l	[127]+
1	<	m	[128]+
1	<	h	[14B]+
i	<	a	[13A]+

$$
\begin{array}{rcl}
c & < & k \\
k & < & n \\
m & < & l \\
p & < & o \\
g & < & b \\
c & < & e \\
a\,l & < & i\,d \\
h\,j & < & b\,p \\
m\,g & < & e\,o \\
(o-1)(d-g) & < & (l-o)(g-1) \\
(b+a)(i+n) & < & (a-f)(i+j) \\
(e-h)(i+m) & < & (e+a)(m-p) \\
(d-c)(j-k) & < & (c-b)(k-l)
\end{array}
\qquad
\begin{array}{l}
[16\text{A}]- \\
[269]+ \\
[278]- \\
[20\text{B}]- \\
[450]- \\
[468]- \\
[37\text{A}]- \\
[5\text{AB}]+ \\
[80\text{A}]- \\
[170]+ \\
[359]- \\
[38\text{B}]+ \\
[567]+
\end{array}
$$

Table 3-2. Inequality system in 16 *positive* variables describing all coordinatizations (modulo projective equivalence) of the oriented matroid χ_{12}.

After computer preparation up to this point, the system in Table 3-2 has been solved (in hard work !!) by D.Ljubić. (At this stage interactive computer algebra methods might be helpful.) The homogenous coordinates which he found for a realization of the oriented matroid χ_{12} are listed in Table 3-3. Due to its special structure, the realization space $\mathcal{R}(\chi_{12})$, see Section 5.1, of this oriented matroid is extremely "narrow", and, in spite of computer graphic support, it was impossible for us to obtain a reasonable diagram for the straight arrangement $\mathcal{A}_{12}$ obtained from Table 3-3.

$$
\begin{pmatrix}
-1 & 1 & 5000 & 0 & 60 & 80 & 160 & 101 & -1 & -15 & 0 & -12 \\
-1 & 0 & 500 & 1 & 90 & 81 & 15 & 12 & 82 & -2 & 0 & -19 \\
1 & 0 & 1 & 0 & -1 & -1 & -1 & -1 & -1 & 1 & 1 & 10
\end{pmatrix}
$$

Table 3-3. Coordinate matrix for the oriented matroid χ_{12}.

In view of the isotopy problem (see Chapter VI) it would be interesting to know whether the semi-algebraic variety in $\mathbf{R}^{16}$ defined by the system in Table 3-2 is connected.

As a second example we consider the affine rank 3 oriented matroid D_3^{10} associated with the pseudo-configuration in Figure 3-13. In that diagram the points $4, 5$ and 6 are supposed to be located "very close" to the intersection points $\overline{12} \wedge \overline{37}$, $\overline{23} \wedge \overline{18}$, and $\overline{13} \wedge \overline{29}$ respectively.

This oriented matroid has been constructed by J.Richter, and it can be shown by an easy geometric argument based on Desargues' theorem that D_3^{10} is not realizable. In order to motivate the following chapter, however, we will present a systematic algebraic non-realizability proof.

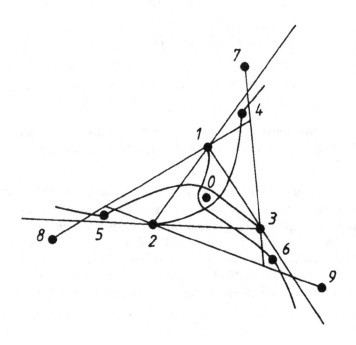

Figure 3-13. The non-realizable oriented matroid D_3^{10}.

First recall the following special cases of the Grassmann-Plücker relations, see Theorem 1.8 and Remark 3.11.

Corollary 3.13. *For any field K and $x, y, z \in K^3$ abbreviate the determinant $\det(x, y, z)$ by the bracket $[xyz]$. Then for all $a, b, c, d, e, f \in K^3$ we have the identities*

$$\{a|bcde\} \quad = \quad [abc][ade] - [abd][ace] + [abe][acd] \quad = \quad 0$$

$$< abc|def > := [abc][abc][def] \ - \ \det \begin{pmatrix} [dbc] & [adc] & [abd] \\ [ebc] & [aec] & [abe] \\ [fbc] & [afc] & [abf] \end{pmatrix} = 0.$$

Assume that there exist points $x_1, x_2, \ldots, x_{10}$ in the real Euclidean plane such that for all i, j, k the oriented volume $[ijk]$ of the triangle x_i, x_j, x_k has the sign prescribed in Figure 3-13. The following expression vanishes by Corollary 3.13.

$\qquad$ $\{1|2345\}$ $[234][153][126][126][230][137][183][293]$

$-\quad$ $\{2|1340\}$ $[234][153][153][126][126][137][183][293]$

$+\quad$ $\{1|2360\}$ $[234][134][152][152][263][137][183][293]$

$-\quad$ $\{3|1246\}$ $[134][152][152][263][120][137][183][293]$

$-\quad$ $\{2|1356\}$ $[234][134][152][126][103][137][183][293]$

$+\quad$ $\{3|1250\}$ $[234][134][152][126][126][137][183][293]$

$-\quad$ $\{1|2358\}$ $[134][263][137][293][120]\,(\,[234][153][126] + [134][152][263]\,)$

$+\quad$ $\{2|1369\}$ $[234][153][137][183][120]\,(\,[234][153][126] + [134][152][263]\,)$

$-\quad$ $\{3|1247\}$ $[153][263][183][129][120]\,(\,[234][153][126] + [134][152][263]\,)$

$+\quad$ $< 123|789 > [134][153][263][120]\,(\,[234][153][126] + [134][152][263]\,)$

By expanding the Grassmann-Plücker terms $\{\ldots|\ldots\}$ and $< \ldots|\ldots >$ in the above polynomial, we obtain after cancelling several summands

$\qquad$ $[123][145][234][153][126][126][230][137][183][293]$

$+\quad$ $[123][240][234][153][153][126][126][137][183][293]$

$+\quad$ $[123][160][234][134][152][152][263][137][183][293]$

$+\quad$ $[123][364][134][152][152][263][120][137][183][293]$

$+\quad$ $[123][256][234][134][152][126][103][137][183][293]$

$+\quad$ $[123][350][234][134][152][126][126][137][183][293]$

$+\quad$ $[123][296][234][153][137][183][120]\,(\,[234][153][126] + [134][152][263]\,)$

$+\quad$ $[123][374][153][263][183][129][120]\,(\,[234][153][126] + [134][152][263]\,)$

$+\quad$ $[123][185][134][263][137][293][120]\,(\,[234][153][126] + [134][152][263]\,)$

$+\quad$ $(\,[123]^2[789] + [172][183][293] + [172][283][139] + [182][237][139]$
$\qquad + [129][137][283]\,) \cdot [134][153][263][120]\,(\,[234][153][126] + [134][152][263]\,)$

All brackets occurring in this polynomial, e.g. $[123]$, $[145], \ldots, [374], [153], \ldots$, have to be positive in any realization of D_3^{10} as is seen from Figure 3-13. Since the sum of 10 positive numbers cannot vanish, it follows that D_3^{10} is not realizable.

Such a polynomial which "obviously" shows the non-realizability of an oriented matroid is called a *final polynomial*, a notion that will be made precise in the next chapter. Here it is our main goal to describe the steps that led to the construction of the above final polynomial.

For that purpose we would like to forget the above proof, and we assume at this point that D_3^{10} is realizable. Then there exists a real 3×10-matrix coordinate matrix A for D_3^{10} with

$$A = \begin{pmatrix} 1 & 0 & 0 & a & d & -g & j & -m & -p & s \\ 0 & 1 & 0 & -b & e & h & -k & n & -q & t \\ 0 & 0 & 1 & c & -f & i & -l & -o & r & u \end{pmatrix}.$$

In Figure 3-13 we see that all variables $a, b, \ldots, u$ have to be positive. For example, we have $[124] = c > 0$ for the oriented volume $[124]$ of the triangle $1, 2, 4$. With Algorithm 3.12 it can easily be shown that the positivity of all 21 variables together with the following 10 inequalities forms a reduced system for D_3^{10}, i.e. D_3^{10} is realizable if and only if the inequality system (1) – (10) has a solution within the positive real numbers.

$$
\begin{aligned}
[123][145] &= bf - ce &> 0 \qquad &(1) \\
[123][185] &= eo - fn &> 0 \qquad &(2) \\
[123][160] &= hu - it &> 0 \qquad &(3) \\
[123][240] &= cs - au &> 0 \qquad &(4) \\
[123][256] &= fg - di &> 0 \qquad &(5) \\
[123][296] &= ip - gr &> 0 \qquad &(6) \\
[123][364] &= bg - ah &> 0 \qquad &(7) \\
[123][374] &= ak - bj &> 0 \qquad &(8) \\
[123][350] &= dt - es &> 0 \qquad &(9) \\
[123]^2[789] &= rjn - lnp - lmq - ojp - okp - rkm &> 0 \qquad &(10)
\end{aligned}
$$

In order to decide the realizability of D_3^{10}, we could proceed from this point on by naive variable elimination. We consider, for example, the variable c which is contained only in (1) and in (4). All variables being positive, these two inequalities can be rewritten as

$$\frac{au}{s} < c < \frac{bf}{e}.$$

Hence a real number c satisfying (1) and (4) exists if and only if

$$bfs - aeu > 0, \qquad (11)$$

and the system (1) – (10) can be replaced by the nine inequalities (2), (3), (5), (6), (7), (8), (9), (10) and (11) in one less variable.

Such elementary solving techniques can be used to finally derive $0 < 0$ which shows the contradiction. In order to obtain a final polynomial from this derivation, we mimic the steps which led to new polynomials as in (11) by forming positive linear combinations of vanishing polynomials. Therefore, we consider all inequalities as (vanishing) Grassmann-Plücker identities under the additional condition that all occurring brackets are positive.

$$
\begin{aligned}
p_1 &= [123][145] - bf + ce \\
p_2 &= [123][185] - eo + fn \\
p_3 &= [123][160] - hu + it \\
p_4 &= [123][240] - cs + au \\
p_5 &= [123][256] - fg + di \\
p_6 &= [123][296] - ip + gr \\
p_7 &= [123][364] - bg + ah \\
p_8 &= [123][374] - ak + bj \\
p_9 &= [123][350] - dt + es \\
p_{10} &= [123][123][789] - rjn + lnp + lmq + ojp + okp + rkm
\end{aligned}
$$

Solving for c as in the derivation of (11) yields the new polynomial

$$
p_{11} := s\,p_1 + e\,p_4 = [123][145]s + [123][240]e - bfs + aue
$$

Solving for h yields

$$
p_{12} := a\,p_3 + u\,p_7 = [123][160]a + [123][364]u - bgu + ita
$$

Solving for d yields

$$
p_{13} := t\,p_5 + i\,p_9 = [123][256]t + [123][350]i + eis - fgt
$$

Solving for t yields

$$
p_{14} := fg\,p_{12} + ia\,p_{13} =
$$

$$
[123][160]afg + [123][364]fgu + [123][256]ait + [123][350]aii + aeiis - bfggu
$$

Solving for s yields

$$
p_{15} := aeii\,p_{11} + bf\,p_{14} =
$$

$$
[123][145]aeiis + [123][240]aeeii + [123][160]abffg + [123][364]bffgu +
$$

$$
[123][256]abfit + [123][350]abfii + u(aei - bfg)(aei + bfg)
$$

Partially solving for j yields

$$
p_{16} := rn\,p_8 + b\,p_{10} =
$$

$$[123][374]rn + [123][123][789]b -$$

$$akrn + blnp + blmq + bojp + bokp + brkm$$

Partially solving for r yields

$$p_{17} := g\, p_{16} + akn\, p_6 =$$

$$[123][296]akn + [123][374]grn + [123][123][789]bg -$$

$$aiknp + bglnp + bglmq + bgjop + bgkpo + bgrkm$$

Partially solving for o yields

$$p_{18} := bgkp\, p_2 + e\, p_{17} =$$

$$[123][296]aekn + [123][374]egrn + [123][123][789]beg + [123][185]bgkp +$$

$$beg(lnp + lmq + jop + rkm) - knp(aei - bfg)$$

Finally, solving for the compound expression $(aei - bfg)$ yields

$$p_{19} := u(aei + bfg)p_{18} + knp\, p_{15} =$$

$$[123][145]aeiiknps + [123][240]aeeiiknp + [123][160]abffgknp +$$

$$[123][364]bffgknpu + [123][256]abfiknpt + [123][350]abfiiknp +$$

$$[123][185]bgkpu(aei + bfg) + [123][296]aeknu(aei + bfg) + [123][374]$$

$$\cdot egrnu(aei + bfg) + (\,[123][123][789] + lnp + lmq + jop + rkm\,)\,begu\,(aei + bfg)$$

To see that p_{19} equals the above final polynomial, we replace all variables by brackets according to the matrix A, e.g. a by [234], b by [134], ... etc. Under these substitutions the expressions $p_1, p_2, \ldots, p_{10}$ equal the syzygy coefficients in the first representation of the final polynomial, e.g. $p_1 = \{1|2345\}$, $p_2 = -\{2|1340\}$, which completes the argument.

Chapter IV

ALGEBRAIC CRITERIA FOR GEOMETRIC REALIZABILITY

This chapter is devoted to the algebraic geometry of matroids and oriented matroids. We develop the theory of final polynomials as a systematic approach to prove non-realizability in computational geometry. This computational method has been introduced by J. Bokowski (see [5], [24], [32]) for constructing short non-polytopality proofs for spheres. In these papers the final polynomials were presented without derivation, and a subsequent joint paper with J. Richter [28a] deals with the derivation and related algorithmic aspects of final polynomials. In view of these results J. Bokowski raised the question of whether a final polynomial exists in every non-realizable case.

It is the main purpose of this chapter to prove an affirmative answer to this question, to discuss various algebraic ramifications, and to present several old and new geometric examples. This result of B. Sturmfels (existence of final polynomials in every non-realizable case) also follows from earlier unpublished results of A. Dress on that subject.

In Section 4.4 we relate our approach to the recent work of Gelfand et.al. [68] on the stratification induced by matroids of the complex Grassmann variety $G_{n,d}^{\mathbf{C}}$. This connection leads to the resolution of a problem of N.White [163] on weak maps and specialization maps of matroid coordinatizations.

4.1. Preliminaries from algebraic and semi-algebraic geometry.

We have seen that the realizability problems for matroids and oriented matroids can be expressed in terms of polynomial equations and inequalities, and that, moreover, arbitrary "bad" polynomials can occur from these problems. In order to study the solution sets of polynomial equations (over arbitrary fields) and inequalities (over ordered fields), we need to employ some tools from classical algebraic geometry and the more recent semi-algebraic geometry.

Here we aim to present some preliminaries from these areas in a relatively self-contained exposition. The first half of the section contains mostly standard material, roughly summarizing the first chapter of Kunz [98]. In the second half we present a real version of Hilbert's Nullstellensatz due to G.Stengle [143] and a semi-algebraic analogue from the survey of E.Becker [12]. For additional algebraic background the reader is refered to e.g. Atiyah & MacDonald [8] and Jacobson [88].

Let K be a subfield of a field L. A subset $V \subset L^n$ is an *(affine algebraic)* K-*variety* if there are polynomials $f_1, \ldots, f_s \in K[x_1, \ldots, x_n]$, such that

$$V \quad = \quad \{\, x \in L^n \mid f_i(x) = 0 \text{ for all } i = 1, \ldots, s \,\}.$$

We write $< f_1, \ldots, f_s >$ for the ideal in $K[x_1, \ldots, x_n]$ generated by $f_1, \ldots, f_s$, and $\sqrt{I}$ for the *radical* of an ideal I, that is the set of all $f \in K[x_1, \ldots, x_n]$ such that $f^m \in I$ for some integer m. An ideal I is *reduced* if $I = \sqrt{I}$, and a ring R is *reduced* if R does not contain any nilpotent elements, or equivalently, its zero ideal is reduced. (Throughout this chapter all rings are commutative and have a unit.)

Given any subset $V \subset L^n$, the *ideal of V* in $K[x_1, \ldots, x_n]$ is the set $\mathcal{I}(V)$ of polynomials $f \in K[x_1, \ldots, x_n]$, such that $f(x) = 0$ for all $x \in V$. Conversely, given any ideal $I \subset K[x_1, \ldots, x_n]$, we define the *variety $\mathcal{V}(I)$ of I* in L^n as the set of all common zeros in L^n of polynomials from I. It follows from Hilbert's basis theorem that every such set $\mathcal{V}(I)$ is indeed a K-variety as defined above.

Theorem 4.1. (Hilbert's basis theorem) [82, Theorem 2.3] *For any field K, the polynomial ring $K[x_1, \ldots, x_n]$ is Noetherian, that is, every ideal I in $K[x_1, \ldots, x_n]$ is finitely generated.*

There is a very close connection between the K-varieties $V \subset L^n$ and the ideals of the ring $K[x_1, \ldots, x_n]$, and hence the ideal theory of commutative rings is very important in algebraic geometry. Here are some easy facts.

Observation 4.2.
a) $\mathcal{I}(\emptyset) = < 1 >$, and $\mathcal{I}(L^n) = < 0 >$ if L is infinite.
b) $\mathcal{I}(V) = \sqrt{\mathcal{I}(V)}$ for every subset $V \subset L^n$.
c) $\mathcal{V}(\mathcal{I}(V)) = V$ for every variety $V \subset L^n$.
d) For any two varieties V_1, V_2, we have $V_1 \subset V_2 \iff \mathcal{I}(V_1) \supset \mathcal{I}(V_2)$ where $V_1 = V_2 \iff \mathcal{I}(V_1) = \mathcal{I}(V_2)$.
e) For any two varieties V_1, V_2, we have $\mathcal{I}(V_1 \cup V_2) = \mathcal{I}(V_1) \cap \mathcal{I}(V_2)$, $V_1 \cup V_2 = \mathcal{V}(\mathcal{I}(V_1) \cdot \mathcal{I}(V_2))$, and $V_1 \cap V_2 = \mathcal{V}(\mathcal{I}(V_1) + \mathcal{I}(V_2))$.

These rules together with Theorem 4.1 imply that finite unions and arbitrary intersections of K-varieties in L^n are again K-varieties. Hence the K-varieties form the closed sets of a topology on L^n, the *Zariski topology* on L^n with respect to K. Observe that if L stands for the complex numbers or a subfield thereof, then the Zariski topology on L^n is weaker than the usual real topology on $L^n \subset \mathbf{C}^n$; it is not even Hausdorff. For example, the only closed sets in the Zariski topology on the complex line $K = L = \mathbf{C}^1$ are the finite subsets of $\mathbf{C}^1$ and $\mathbf{C}^1$ itself.

Observation 4.2 shows also that the assignment $V \mapsto \mathcal{I}(V)$ defines an injective, inclusion-reversing map from the set of K-varieties in L^n into the set of reduced ideals in $K[x_1, \ldots, x_n]$. It is easy to see that the map $\mathcal{I}$ is not surjective if L does not contain the algebraic closure $\overline{K}$ of K. For example, let $K = L = \mathbf{Q}$, the rationals, then there is no variety $V \subset \mathbf{Q}^1$ with $\mathcal{I}(V) = < x^2 + 1 >$ in $\mathbf{Q}[x]$. On the other hand, if L is algebraically closed, then the map $\mathcal{I}$ is bijective by Hilbert's Nullstellensatz (= theorem of zeros).

Theorem 4.3. (Hilbert's Nullstellensatz) [82, Theorem 3.7] *Let L/K be a field extension, where L is algebraically closed. Then the map $V \mapsto \mathcal{I}(V)$ defines a bijection from the set of K-varieties in L^n onto the set of reduced ideals in $K[x_1, \ldots, x_n]$. For every ideal $I \subset K[x_1, \ldots, x_n]$, we have $\sqrt{I} = \mathcal{I}(\mathcal{V}(I))$.*

This correspondence between ideals and varieties in the case of algebraically closed fields is the reason why many text books on algebraic geometry start with the sentence "Let K be an algebraically closed field" which, at first sight, might sound unsatisfactory for the novice who is eager to learn something about real or rational solutions to polynomial equations. To see that Hilbert's Nullstellensatz can be interpreted as a "theorem of the alternative" for polynomial systems, let us derive the following

Corollary 4.4. *Let L/K be a field extension, where L is algebraically closed, and let $f_1, \ldots, f_s, g \in K[x_1, \ldots, x_n]$. Then one and only one of the following statements is true.*
  a) *There exists an $x \in L^n$, such that $f_1(x) = \ldots = f_s(x) = 0$ and $g(x) \neq 0$.*
  b) *There exist $h_1, \ldots, h_s \in K[x_1, \ldots, x_n]$, such that $g^m = h_1 f_1 + \ldots + h_s f_s$ for some $m \in \mathbb{N}$.*

Proof. It is clear that a) cannot hold if b) holds. Conversely, assume that b) does not hold. Writing $I := <f_1, \ldots, f_s> \subset K[x_1, \ldots, x_n]$, this is equivalent to $g \notin \sqrt{I}$. Hence, by Theorem 4.3 $g \notin \mathcal{I}(\mathcal{V}(I))$ which is in turn equivalent to the existence of an $x \in \mathcal{V}(I)$ with $g(x) \neq 0$. $\qquad\square$

Corollary 4.4 describes the desirable output of a computer program that solves polynomial equations. Given the input data $(f_1, \ldots, f_s, g)$, we would like such an algorithm to generate either a *primal solution* $x \in L^n$ as in a) or a *dual solution* $(h_1, \ldots, h_s)$ as in b). In the geometric application to be discussed in the next sections, the primal solution describes a realization of the geometric object while the dual solution corresponds to a non-realizability proof.

The *Gröbner basis method* by B.Buchberger [40] has received much attention in the last decade, and today the terms "Gröbner basis" and "computational algebraic geometry" are frequently used synonymously [10]. For our purposes we can think of a Gröbner basis of an ideal $I \subset K[x_1, \ldots, x_n]$ as a finite generating subset $GB(I) = \{g_1, \ldots, g_t\}$ of I which satisfies certain strong computational completeness properties. In particular, it is true for any Gröbner basis $GB(I)$ of I that $1 \in I$ if and only if $1 \in GB(I)$.

It has been shown in [150] that the presently implemented Gröbner basis procedures can be used to find also the desired primal or dual solution along with the output "$1 \in I$" or "$1 \notin I$". Note at this point that it can be assumed without loss of generality in Corollary 4.4 that $g = 1$, i.e. that there are only equations. For, $g \in \sqrt{I}$ in $K[x_1, \ldots, x_n]$ if and only if in the polynomial ring

$K[x_1, \ldots, x_n, y]$, y a new variable, the ideal $I' := I + < g \cdot y - 1 >$ contains 1. The process of replacing an inequality $g \neq 0$ by an equation $g \cdot y - 1 = 0$ is a special case of the general technique of forming fractions in commutative algebra.

Given any ring R and any multiplicatively closed subset $S \subset R$, the *ring of fractions* R with respect to S is the ring $S^{-1}R$ which is defined as follows. Introduce an equivalence relation on the set $R \times S$ by $(r, s) \sim (r', s')$ if and only if there exists an $s'' \in S$ with $s''(rs' - r's) = 0$. Writing $\frac{r}{s}$ for the equivalence class of (r, s), the usual addition and multiplication rules for fractions define the ring structure on the set $S^{-1}R$ of equivalence classes. As an example consider the multiplicative semigroup S generated by a finite number of elements $r_1, \ldots, r_m$ in R, and let $r := r_1 \cdot \ldots \cdot r_m$. Then $S^{-1}R$ is isomorphic to $R[z] / < r \cdot z - 1 >$ where z is a (new) variable [82, Exercise III.4.1].

Given any K-variety $V \subset L^n$, then the *coordinate ring* of V is the quotient

$$K[V] \quad := \quad K[x_1, \ldots, x_n] / \mathcal{I}(V).$$

The elements $\phi \in K[V]$ are in natural one-to-one correspondence with the polynomial functions $\phi : V \to L$ because any two polynomials $p_1, p_2 \in K[x_1, \ldots, x_n]$ agree as functions on V if and only if $p_1 - p_2 \in \mathcal{I}(V)$, that is if their difference is zero in $K[V]$.

For a subset I of $K[V]$ we define the *variety* $\mathcal{V}_V(I)$ of I *in* V, and for every subset $W \subset V$ we define the *ideal* $\mathcal{I}_V(W)$ *in* $K[V]$ in the canonical way. Clearly, we have $\mathcal{I}_V(W) = \mathcal{I}(W)/\mathcal{I}(V)$, and the rules in Observation 4.2 remain valid in this setting. Moreover, we have the following generalized form of Hilbert's Nullstellensatz.

Theorem 4.5. [82, Theorem 3.11] *Let L/K as in Theorem 4.3 and $V \subset L^n$ a K-variety. The map $W \mapsto \mathcal{I}_V(W)$ which assigns to every K-variety $W \subset V$ its ideal in $K[V]$ is an inclusion-reversing bijection from the K-subvarieties of V onto the reduced ideals in $K[V]$, and $\sqrt{I} = \mathcal{I}_V(\mathcal{V}_V(I))$ for every ideal I of $K[V]$.*

Theorem 4.5 implies that we can ignore the ambient space L^n and work with the coordinate ring $K[V]$ while studying the points and subvarieties of a K-variety $V \subset L^n$. Coordinate rings of K-varieties are *affine K-algebras*, that is they are finitely generated and reduced. Conversely, every affine K-algebra is the coordinate ring of a K-variety in some affine space L^n. After defining morphisms as polynomial maps and ring homomorphisms respectively, this can be stated in a very fancy way : *there is a contra-equivalence between the category of K-varieties and the category of affine K-algebras.*

To make this point of view more transparent, let us see how a K-variety $V \subset L^n$ can be reconstructed from its coordinate ring $K[V]$. The points $x \in V$ are in one-to-one correspondence with the ring homomorphisms from $K[V]$ into L via the map

$$\Phi : V \quad \to \quad \mathrm{Hom}(K[V], L)$$
$$x \quad \mapsto \quad \Phi_x$$

with $\Phi_x(f) := f(x)$. Thus, given any affine K-algebra R and any field extension L of K, the K-variety of R in L^n can be "defined" as $\mathrm{Hom}(R, L)$. The kernel of a homomorphism from R into a field extension L of K is a prime ideal, and conversely, every prime ideal p in R induces a homomorphism from R into the field of fractions of the domain R/p (which is a field extension of K).

Now we are ready to generalize and to define a geometric structure on an arbitrary ring R. The *spectrum* of a ring R, denoted $\mathrm{Spec}(R)$, is the set of prime ideals of R. Given any ideal $I \subset R$, the *variety* of I is the set

$$\mathcal{V}(I) \quad := \quad \{\, p \in \mathrm{Spec}(R) \mid p \supset I \,\}.$$

The sets $\mathcal{V}(I)$ form the closed sets of the *Zariski topology* on $\mathrm{Spec}(R)$. The subset $\mathrm{Max}(R)$ of maximal ideals in R inherits the relative topology.

If R is an affine K-algebra and $V \subset \overline{K}^n$ its variety with respect to the algebraic closure of K, then $\mathrm{Max}(R)$ is homeomorphic to V (with the earlier defined Zariski topology). In general, the elements of $\mathrm{Spec}(R)$ correspond to the irreducible K-subvarieties of V. Forming of fractions, subvarieties and other geometric operations can easily be described using the spectrum of a ring.

Remark 4.6. (compare [82, Sect.I.4]) *Let I be any ideal and S any multiplicatively closed set in R. Then we have the following homeomorphisms (in the respective Zariski topologies).*
   a) $\mathrm{Spec}(R/I) = \mathcal{V}(I) \subset \mathrm{Spec}(R)$.
   b) $\mathrm{Spec}(S^{-1}R) = \{\, p \in \mathrm{Spec}(R) \mid p \cap S = \emptyset \,\}$.

Let us apply these abstract concepts to the concrete problem of solving polynomial equations and inequalities. Given any field K, we ask whether the system

$$f_1(x) = \ldots = f_s(x) = 0, \quad g_1(x) \neq 0, \ldots, g_s(x) \neq 0 \qquad (P)$$

with $f_1, \ldots, f_s, g_1, \ldots, g_t \in K[x_1, \ldots, x_n]$ has a solution over some field extension L of K. Let S denote the multiplicative semigroup in $K[x_1, \ldots, x_n]$ generated by $g_1, \ldots, g_s$, and let $I := \langle f_1, \ldots, f_t \rangle$. The topological space $\mathrm{Spec}(S^{-1} \cdot (R/I))$ will be called the *Zariski solution space* to the problem (P).

Proposition 4.7. *The points of the Zariski solution space $\mathrm{Spec}(S^{-1} \cdot (R/I))$ are in one-to-one correspondence with solutions of problem (P) over field extensions of K. This correspondence defines a homeomorphism from the solutions of (P) in $\overline{K}^n$ (in the K-Zariski topology) onto $\mathrm{Max}(S^{-1} \cdot (R/I))$.*

By Remark 4.6 we have

$$\mathrm{Spec}(S^{-1} \cdot (R/I)) \quad = \quad \{\, p \in \mathrm{Spec}(R) \mid p \supset I \text{ and } p \cap S = \emptyset \,\}.$$

So we can rephrase Corollary 4.4 by stating that problem (P) either has a Zariski solution, and hence a solution in $\overline{K}$, or (P) does not have a Zariski solution, that is there is no prime ideal p in R with $I \subset p$ and $p \cap S = \emptyset$. The last condition is equivalent to $I \cap S \neq \emptyset$ by a lemma of Krull ([82, Lemma 4.4]). (Observe that the "dual solution" in Corollary 4.4 b) can be interpreted as an element of $I \cap S$.)

In order to derive the *real* Nullstellensatz of Stengle [124, Theorem 2], let us briefly recall some basics of the *Artin-Schreier theory*. This theory forms the backbone of Artin's affirmative solution to Hilbert's 17th problem whether every polynomial which is positive on $\mathbf{R}^n$ can be written as a sum of squares of rational functions. For details the reader is refered to [72, Chapter VI].

A field K is *formally real* if -1 is not a sum of squares in K. It follows relatively easily that every formally real field K can be ordered and that K is necessarily of characteristic 0. Conversely, every ordered field is formally real. A field L is *real closed* if L is formally real and no proper algebraic extension of L is formally real. Paradigms of real closed fields are the real algebraic numbers $\mathbf{A}$ and the real numbers $\mathbf{R}$.

Here are some more facts, all of which have been proved in [72, Chap.VI]. A field L is real closed if and only if $\sqrt{-1} \notin L$, and $L(\sqrt{-1})$ is algebraically closed. Any ordered field K admits an order-preserving embedding into a unique (up to order isomorphism) real closed field $\tilde{K}$ which is algebraic over K. $\tilde{K}$ is the *real closure* of K. A system of polynomial equations and inequalities with coefficients in some ordered field K has a solution in some ordered field extension L of K if and only if it has a solution in $\tilde{K}$. In particular, such a system with rational coefficients has a solution in some ordered field if and only if it has a solution in the field $\mathbf{A} = \tilde{\mathbf{Q}}$ of real algebraic numbers. This is the famous universality property of the real algebraic numbers that we already mentioned in Section 2.2.

Given an ordered field K, we write

$$K[x_1, \ldots, x_n]^+ \quad := \quad \left\{ \sum \alpha_i f_i^2 \mid \alpha_i \in K^+ \text{ and } f_i \in K[x_1, \ldots, x_n] \right\}$$

for the semi-ring of "obviously non-negative" polynomials. Given any ideal $I \subset K[x_1, \ldots, x_n]$, the *real radical* of I can be defined as

$$\sqrt[R]{I} \quad := \quad \left\{ f \mid f^{2m} \in I - K[x_1, \ldots, x_n]^+ \text{ for some } m \in \mathbf{N} \right\}.$$

Then we have the following real Nullstellensatz due to G.Stengle. Let us remark that an earlier variant of this result had been obtained in the late sixties by D.W. Dubois [60].

Theorem 4.8. (Stengle [124, Theorem 2]) *Let L/K be an ordered field extension, where L is real closed. Then for every ideal $I \subset K[x_1, \ldots, x_n]$ we have $\sqrt[R]{I} = \mathcal{I}(\mathcal{V}(I))$.*

Also Stengle's Nullstellensatz can be interpreted as a "theorem of the alternative" describing the primal and the dual solution in a polynomial programming problem. The proof of Corollary 4.9 is immediate.

Corollary 4.9. *Let L/K be an ordered field extension, where L is real closed, and let $f_1, \ldots, f_s, g \in K[x_1, \ldots, x_n]$. Then one and only one of the following statements is true.*

   a) *There exists an $x \in L^n$ such that $f_1(x) = \ldots = f_s(x) = 0$ and $g(x) \neq 0$.*
   b) *There exist $p_1, \ldots, p_s, q_1, \ldots, q_t \in K[x_1, \ldots, x_n]$, $\alpha_1, \ldots, \alpha_t \in K^+$ and $m \in \mathbb{N}$ such that*

$$g^{2m} + p_1 f_1 + \ldots + p_s f_s + \alpha_1 q_1^2 + \ldots + \alpha_t q_t^2 = 0$$

We have seen above that the spectrum $\mathrm{Spec}(R)$ of an affine K-algebra R corresponds to the (irreducible) K-subvarieties of the variety $V := \mathrm{Max}(R) \hookrightarrow \overline{K}^n$ associated with R, and that the "points" of V can be thought of as homomorphisms from R into $\overline{K}$. If K is ordered, then we are especially interested in the homomorphisms from R into ordered fields. This suggests to us to introduce the *real spectrum* $\mathrm{Rspec}(R)$ and the *maximal real spectrum* corresponding to homomorphisms from R into the real closure $\tilde{K}$ of K.

The following intrinsic definition of the real spectrum for an arbitrary ring R has been given by E. Becker [9, Section 2]. A subset $J \subset R$ is a *prime ordering* if

$$J + J \subset J, \quad J \cdot J \subset J, \quad J \cup -J = R, \quad J \cap -J \text{ is a prime ideal.}$$

Becker defines $\mathrm{Rspec}(R)$ as the set of prime orderings of R, and he introduces a natural (real) topology on the real spectrum which is finer than the relative topology induced by the injective map $\mathrm{Rspec}(R) \to \mathrm{Spec}(R)$, $J \mapsto J \cap -J$.

The set-up of the real spectrum is used in [9, Section 4] to prove various powerful semi-algebraic Nullstellen- and Positivstellensätze. The following result, for example, is a generalization of Stengle's Nullstellensatz (Theorem 4.8).

Theorem 4.10. [9, Theorem 4.2] *Given any ring R, then $\mathrm{Rspec}(R) = \emptyset$ if and only if $-1 \notin \sum R^2$.*

Theorem 4.10 states in other words that a ring R is formally real if and only if it does not admit any non-trivial homomorphism into an ordered field.

Let K be a subfield of a real closed field L, and let V be a K-variety in L^n with coordinate ring $K[V]$. A subset W of V is a *semi-algebraic K-variety* in V if there are $f_1, \ldots, f_r, g_1, \ldots, g_s, h_1, \ldots, h_t \in K[x_1, \ldots, x_n]$ such that $W = \{ x \in V \mid f_1(x) = \ldots = f_r(x) = 0, g_1(x) \geq 0, \ldots, g_s(x) \geq 0, h_1(x) > 0, \ldots, h_t(x) > 0 \}$.

The following semi-algebraic Nullstellensatz characterizes the set of all polynomials in $K[V]$ which vanish on W.

Let H be the multiplicative semi-group with unit in $K[V]$ generated by $h_1, \ldots, h_t$, and let T be the quadratic semi-ring in $K[V]$ generated by $K[V]^2$, K^+, and the functions $g_1, \ldots, g_s, h_1, \ldots, h_t$. Finally, abbreviate $I := \langle f_1, \ldots, f_r \rangle \subset K[V]$.

Theorem 4.11. (Becker, [9, Theorem 4.7]) *The following statements are equivalent.*
  a) $f(x) = 0$ for all $x \in W$.
  b) $h \cdot f^{2m} \in I - T$ for some $h \in H$ and $m \in \mathbf{N}$.

This statement can again be easily translated into a theorem of the alternative for polynomial programming. In setting $f = 1$ in Theorem 4.11 we obtain

Corollary 4.12. *With the notations and hypotheses as above, either the semi-algebraic variety W is non-empty, or there exists a polynomial $p \in (H + T) \cap I$.*

Let us close this section by remarking that the currently implemented decision routines for the elementary theory of real closed fields, such as Collins' Cylindrial Algebraic Decomposition [43], cannot be modified to generate either a primal solution $x \in W$ or a dual solution $p \in (H + T) \cap I$. Indeed, it has been pointed out by M.F. Coste-Roy [private communication] that presently no *constructive* version of Theorem 4.11 is known in real algebraic geometry.

4.2. Final polynomials for matroids

The concepts introduced in the previous section will be applied in the following to the realizability of matroids. Let us outline the basic ideas. Given a field K, we shall assign to every rank d matroid M on $E = \{1, 2, \ldots, n\}$ an ideal I_M^K and a multiplicatively closed subset S_M^K of the coordinate ring $K[G_{n,d}]$ of the Grassmann variety $G_{n,d}^K$. The ring $R_M^K := S_M^{K^{-1}} \cdot (K[G_{n,d}]/I_M^K)$ will be called the K-*coordinate ring* of M because the points in $\mathrm{Spec}(R_M^K)$ are in one-to-one correspondence with the realizations of M over field extensions of K (modulo special linear group). Moreover, if K is ordered, then the real spectrum $\mathrm{Rspec}(R_M^K)$ corresponds to realizations of M over ordered extensions of K.

In view of Bokowski's results in [24], [32], we used to think of *final polynomials* as bracket polynomials which yield an "obvious" non-realizability proof for a geometric structure. However, not being aware of the underlying algebraic geometry, it was not immediately clear to us what the most suitable definition for such polynomials should be.

In this section we derive definitions for final polynomials with respect to arbitrary fields and ordered fields which are "correct" in the sense that final polynomials exist in every non-realizable instance. We prove these existence theorems, and we discuss two non-trivial geometric applications of the final polynomial method.

Our methods are closely related to the work of N.White [160] on the *bracket ring* and N.Fenton [63] on the *Vamos ring* associated with a matroid. Both authors consider their rings with integer coefficients, but as White points out in [163], the transition between integer coefficients and coefficients from some field is straightforward. Let us write $\mathbf{Z}[G_{n,d}]$ for the analogous Grassmann ring (brackets modulo Grassmann-Plücker syzygies) with integer coeffients. White's *bracket ring* of M is the ring $B_M := \mathbf{Z}[G_{n,d}]/I_M^{\mathbf{Z}}$, and $\mathrm{Spec}(B_M)$ is the space of all weak coordinatizations of M. In particular, the bracket ring B_U^K with coefficients in some field K of the rank d uniform matroid $U_{n,d}$ on n elements is the coordinate ring $K[G_{n,d}]$ of the Grassmann variety.

Our coordinate ring $R_M^{\mathbf{Z}}$ with integer coefficients is isomorphic to Fenton's *simplified Vamos ring* R_M [50, Sect.2]. Compared to White's approach Fenton goes further, he forms fractions in order to get rid of the annoying degeneracies. Another difference to White's work is that Fenton defines his rings in terms of bases, proving the basis-independence of his definition afterwards. Under that aspect we prefer White's basis-independent bracket ring, and our definitions attempt to combine the advantages of both of the earlier directions. Although the authors have to admit that all computations of the specific final polynomials in this chapter have been carried out with respect to some basis, the basis free bracket version seems much more appropriate for presenting our practical as well as theoretical results (compare Section 3.3).

Recall the following definitions from Section 1.2. Given integers $n \geq d \geq 1$ and a field K, let $K[\Lambda(n, d)]$ denote the ring which is freely generated over K by all brackets $[\lambda]$, $\lambda \in \Lambda(n, d)$. As before, we write $[\lambda_{\pi(1)} \ldots \lambda_{\pi(d)}] := \operatorname{sign} \pi \cdot [\lambda_1 \ldots \lambda_d]$ for any permutation π. Via Plücker coordinates we view $K[\Lambda(n, d)]$ as the ring of polynomial functions on the $\binom{n}{d}$–dimensional vector space $\wedge_d K^n$. Write $G_{n,d}^K \subset \wedge_d K^n$ for the Grassmann variety of simple d-vectors in $\wedge_d K^n$.

Let $I_{n,d}^K$ denote the ideal generated in $K[\Lambda(n, d)]$ by all quadratic syzygies

$$\sum_{i=1}^{d+1} (-1)^i \cdot [\lambda_1, \ldots, \lambda_{i-1}, \lambda_{i+1}, \ldots, \lambda_{d+1}] \cdot [\lambda_i, \mu_1, \ldots, \mu_{d-1}],$$

where $\lambda \in \Lambda(n, d+1)$, $\mu \in \Lambda(n, d-1)$. In the terminology of the last section Theorem 1.8 can be reformulated as

Corollary 4.13. *For every infinite field K we have*

$$\mathcal{I}(\mathcal{V}(I_{n,d}^K)) \quad = \quad \mathcal{I}(G_{n,d}^K) \quad = \quad I_{n,d}^K,$$

and so the points of the Grassmann variety $G_{n,d}$ can be identified with the homomorphisms from its coordinate ring $K[G_{n,d}] := K[\Lambda(n, d)]/I_{n,d}^K$ into K.

Let M be a rank d matroid on $E = \{1, 2, \ldots, n\}$, and let L/K be a field extension. A point in $G_{n,d}^L$, that is a homomorphism $\phi : K[G_{n,d}] \to L$, corresponds to a realization of M over L provided $\phi([\lambda]) \neq 0$ if and only if λ is a basis of M for all $\lambda \in \Lambda(n, d)$. Likewise ϕ corresponds to a weak realization of M provided $\phi([\lambda]) \neq 0$ only if λ is a basis of M.

We assign to M the two sets of all bracket polynomials which must (resp. cannot) vanish under a realization homomorphism ϕ as follows. Let I_M^K denote the ideal in $K[G_{n,d}]$ which is generated by

$$\{ [\lambda], \lambda \in \Lambda(n, d) \text{ is dependent in } M \},$$

and let S_M^K denote the multiplicative semi-group with unit generated by

$$\{ [\lambda], \lambda \text{ is a basis of } M \}.$$

In other words, I_M^K consists of all linear combinations of non-basis brackets with polynomial coefficients, and S_M^K consists of all non-zero monomials which are products of basis brackets. The map $\phi : K[G_{n,d}] \to L$ is a weak realization of M if and only if $I_M^K \subset \operatorname{Ker} \phi$, and ϕ is a realization if, in addition, $S_M^K \cap \operatorname{Ker} \phi = \emptyset$.

By standard abuse of notation we write I_M^K and S_M^K also for the corresponding sets in the free polynomial algebra $K[\Lambda(n, d)]$. Since all computations are most conveniently carried out in $K[\Lambda(n, d)]$, we rephrase the above observation.

Remark 4.14. *Let M be a matroid as above. The coordinatizations of M over L (modulo special linear group) are in one-to-one correspondence with the ring homomorphisms $\phi : K[\Lambda(n,d)] \to L$ such that $I_{n,d}^K + I_M^K \subset \mathrm{Ker}\,\phi$ and $S_M^K \cap \mathrm{Ker}\,\phi = \emptyset$.*

Therefore we define the *K-coordinate ring* of M as

$$R_M^K \quad := \quad {S_M^K}^{-1} \cdot (K[G_{n,d}]/I_M^K) \quad = \quad {S_M^K}^{-1} \cdot (K[\Lambda(n,d)]/(I_{n,d}^K + I_M^K)).$$

$\mathrm{Spec}(R_M^K)$ is the space of all realizations over field extensions of K, and by Proposition 4.7, $\mathrm{Max}(R_M^K)$ is the space of all realizations of M over $\overline{K}$.

A *final polynomial* for M (with coefficients in K) is a bracket polynomial $f \in K[\Lambda(n,d)]$ such that $f \in I_{n,d}^K \cap (S_M^K + I_M^K)$. Since all polynomials defining $I_{n,d}^K, S_M^K$ and I_M^K have integer coeffients, there is a final polynomial with coefficients in K if and only if there is such a polynomial with coefficients in K', the prime field of K. Therefore, we can assume without loss of generality that K is a prime field.

It follows directly from the definitions that there exists a final polynomial for f with coefficients in K if and only if $\mathrm{Spec}(R_M^K)$ is empty. The same argument remains valid if we replace K by the ring of integers $\mathbf{Z}$.

Proposition 4.15.

 a) *Let K be the prime field of characteristic q, $q = 0$ or q prime. There exists a final polynomial for M with coefficients in K if and only if M is not realizable over any field of characteristic q.*

 b) *There exists a final polynomial for M with coefficients in $\mathbf{Z}$ if and only if M is not realizable over any field.*

Let us note that part b) follows also from the results of Fenton [50, Theorem 1.2)] and White [139, Corollary 4.7]. Observe also that the proof of Proposition 4.15 is based only on relatively simple ring-theoretic arguments. Hilbert's Nullstellensatz is not needed to prove Proposition 4.15 or Fenton's and White's analogues !

The desired stronger "theorem of the alternative" for realizabilty of matroids follows in a straightforward manner from Corollary 4.4, which was derived as an equivalent version from Hilbert's Nullstellensatz.

Theorem 4.16. *Let K be the prime field of characteristic q and $\overline{K}$ its algebraic closure. There exists a final polynomial for M with coefficients in K if and only if M is not realizable over $\overline{K}$.*

As pointed out earlier, we consider characteristic 0 as the most interesting case.

Corollary 4.17. *A matroid M is either realizable over the complex numbers* **C** *or there is a final polynomial with rational coefficients for M.*

Example 4.18. *A non-realizable torus.*

Consider the 2-dimensional cell complex C with 9 vertices 19 edges, and 10 facets, 8 quadrangles and 2 triangles depicted in Figure 4-1.

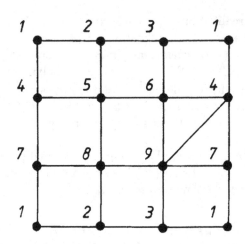

Figure 4-1. The "non-realizable" torus C with 9 vertices.

It is well-known and easy to see that C can be embedded into Euclidean 3-space with all eight 4-sided facets as flat quadrangles, and such that the two adjacent triangular faces 469 and 479 are in one plane. Is there a *proper* embedding of C, i.e. such that 469 and 479 are not coplanar ?

Simutis [140] has shown that a proper embedding of C cannot exist with convex faces, and recently B.Grünbaum [private communication] raised the question whether C admits a proper embedding with possibly non-convex quadrangular faces.

Here we use the final polynomial method to prove that such an embedding does not exist. A geometric non-realizability proof has been obtained simultaneously and independently by D.Ljubić, see also [107] for a related result of his.

Consider the rank 4 matroid M_C on $E = \{1, 2, \ldots, 9\}$ defined by its non-bases

[1245], [1278], [1346], [1379], [2356], [2389], [4578], [5689].

We prove that M_C is not realizable over any field K by constructing a final polynomial with integer coefficients. This will imply in particular that there is no proper emdedding of C into $\mathbf{R}^3$.

As before in Section 3.3, we write $\{\,\sigma\,|\,\tau\,\}$ for the three-term syzygies in $I_{n,d}^K$. Furthermore, we introduce the abbreviation $<\beta_1 \ldots \beta_d\,|\,\lambda_1 \ldots \lambda_d> :=$

$$[\lambda]\cdot[\beta]^{k-1} - [\beta]^{k-d}\cdot\det\begin{pmatrix}[\lambda_1\beta_2\ldots\beta_d] & [\beta_1\lambda_1\ldots\beta_d] & \ldots & [\beta_1\beta_2\ldots\lambda_1]\\ [\lambda_2\beta_2\ldots\beta_d] & [\beta_1\lambda_2\ldots\beta_d] & \ldots & [\beta_1\beta_2\ldots\lambda_2]\\ \vdots & \vdots & \ddots & \vdots\\ [\lambda_d\beta_2\ldots\beta_d] & [\beta_1\lambda_d\ldots\beta_d] & \ldots & [\beta_1\beta_2\ldots\lambda_d]\end{pmatrix}$$

where $\lambda, \beta \in \Lambda(n,d)$ and $k := |\lambda\setminus\beta|$. Observe that the $d\times d$-determinant reduces to a $k\times k$-determinant and that $<\beta\,|\,\lambda>$ is one of the *van der Waerden identities* in [139, Section 6]. In particular, $<\beta\,|\,\lambda>$ is contained in the Grassmann ideal $I_{n,d}^K$.

Now consider the polynomial $f_C \in \mathbf{Z}[\Lambda(9,4)]$ which is given by
$f_C :=$

$\quad - \{18|2736\}[5368][1348][1365][1367][1398]\cdot([4368][1968] - [1468][9368])$

$\quad + \{13|7968\}[5368][1328][1365]\cdot([4368][1968] - [1468][9368])$

$\qquad\qquad\qquad \cdot([1348][1768] - [1468][1378])$

$\quad + \{36|2518\}[1367][1398]\cdot([4368][1968] - [1468][9368])$

$\qquad\qquad\qquad \cdot([1468][1358][1378] - [1348][1568][1378])$

$\quad + \{38|2916\}[1365][1378][1367]\cdot([4368][1968] - [1468][9368])$

$\qquad\qquad\qquad \cdot([1348][1568] - [1468][1358])$

$\quad - \{68|5913\}[1328][4368][1365][1378][1369]\cdot([1348][1768] - [1468][1378])$

$\quad + <1368|1245>[5368][1378][1367][1398]\cdot([4368][1968] - [1468][9368])$

$\quad + <1368|4578>[1328][1365][1378]\cdot([1468][1358][9368] - [1348][1568][9368])$

$\quad + <1368|4768>[1328][1468][1365][1378][9368][1369].$

The polynomial f_C is obviously contained in $I_{9,4}^{\mathbf{Z}}$. On the other hand, we obtain by expansion of the terms $\{\ldots|\ldots\}$ and $<\ldots|\ldots>$ the following representation.

$$f_C \ =$$

$- \underline{[1278]}[1368][5368][1348][1365][1367][1398] \cdot ([4368][1968] - [1468][9368])$

$+ \underline{[1379]}[1368][5368][1328][1365] \cdot ([4368][1968] - [1468][9368])$

$\qquad\qquad \cdot ([1348][1768] - [1468][1378])$

$+ \underline{[2365]}[1368][1367][1398] \cdot ([4368][1968] - [1468][9368])$

$\qquad\qquad \cdot ([1468][1358][1378] - [1348][1568][1378])$

$+ \underline{[2398]}[1368][1365][1378][1367] \cdot ([4368][1968] - [1468][9368])$

$\qquad\qquad \cdot ([1348][1568] - [1468][1358])$

$- \underline{[5968]}[1368][1328][4368][1365][1378][1369] \cdot ([1348][1768] - [1468][1378])$

$+ \{ \underline{[1245]}[1368][1368] + \underline{[1364]} ([1268][1358] - [1328][1568])$

$\qquad\qquad \cdot [5368][1378][1367][1398] \cdot ([4368][1968] - [1468][9368])$

$+ \{ \underline{[4578]}[1368][1368] - \underline{[1364]} ([1768][9368] - [7368][1968])$

$\qquad\qquad \cdot [1328][1365][1378] \cdot ([1468][1358][9368] - [1348][1568][9368])$

$+ [4769][1368][1368][1328][1468][1365][1378][9368][1369]$

This representation shows that f_C is contained in $I_{M_C}^{\mathbf{Z}} + S_{M_C}^{\mathbf{Z}}$ because a bracket $[ijkl]$ corresponds to a basis of M_C if and only if it is not underlined. Therefore, f_C is a final polynomial for M_C with integer coeffients, and hence M_C is not realizable over any field by Proposition 4.15. $\qquad\qquad\square$

Let us now come to the case of an ordered field K. As in Section 4.2, we write $K[\Lambda(n,d)]^+$ for the semi-ring which is generated by the squares in $K[\Lambda(n,d)]$ and the positive field elements K^+. Note that if K is real closed, then every element of K^+ is a square in K, which is clearly false for the rationals $K = \mathbf{Q}$. Given a rank d matroid M on $E = \{1, 2, \ldots, n\}$, we define a *real final polynomial* for M (with coefficients in K) to be a bracket polynomial $f \in K[\Lambda(n,d)]$ such that

$$f \ \in \ I_{n,d}^K \cap \big(I_M^K + K[\Lambda(n,d)]^+ + (S_M^K)^2 \big).$$

As before, we can restrict ourselves to the case of a prime field. Since there is no ordered prime field but the rationals, there is no loss of generality if we state the result only for $K = \mathbf{Q}$.

Theorem 4.19. *The following statements are equivalent for a matroid M.*
(a) There exists a real final polynomial for M with rational coefficients.
(b) M cannot be coordinatized over any ordered field.
(c) M cannot be coordinatized over the real algebraic numbers $\widetilde{\mathbf{Q}} = \mathbf{A}$, the real closure of $\mathbf{Q}$.

Proof. According to the results of the previous section, the realizations of M over ordered field extensions of $\mathbf{Q}$, that is the realizations of M over any ordered field, are in one-to-one correspondence with the elements of the real spectrum $\mathrm{Rspec}\,(R_M^{\mathbf{Q}})$ of the $\mathbf{Q}$-coordinate ring of M. Now the proof follows from the fact that by Theorem 4.8, Corollary 4.9 and Theorem 4.10 also the conditions (a) and (b) are equivalent to $\mathrm{Rspec}\,(R_M^{\mathbf{Q}}) = \emptyset$.

$\square$

Let us see an example.

Example 4.20. *The complex configuration 8_3.*

We have discussed this configuration earlier in Section 3.1, see also Figure 3-1. The matroid $M = M_{8_3}$ is the rank 3 matroid on $E = \{1, 2, \ldots, 8\}$, given by its non-bases

$$[124], [235], [346], [457], [568], [671], [782], [813].$$

Consider the following polynomial $f_{8_3} \quad :=$

$+ \ [257][354][317][237][351] \cdot \{5/3786\}$

$+ \ [257][354][317][356][351] \cdot \{7/3582\}$

$+ \ [257][354][317][356][257] \cdot \{3/5781\}$

$+ \ [257][354][237][351][358] \cdot \{7/3561\}$

$+ \ [257][157][351][237][358] \cdot \{3/5764\}$

$+ \ [157][327][356][358][351] \cdot \{7/5324\}$

$+ \ [157][327][356][358][354] \cdot \{7/5312\}$

$+ \ [157][327][356][358][357] \cdot (-[357][124] + [351][724] - [352][714] + [354][712])$

which is clearly contained in $I_M^{\mathbf{Q}} \subset \mathbf{Q}[\Lambda(8,3)]$.

By expanding the quadratic Grassmann-Plücker syzygies, we obtain after cancellation of several summands the representation

$$
\begin{aligned}
f_{8_3} \quad = & - \ [157][327][356][358][357][357]\ \underline{[124]} \\
& - \ [157][327][356][358][357][714]\ \underline{[235]} \\
& - \ [257][157][351][237][358][357]\ \underline{[346]} \\
& - \ [157][327][356][358][351][732]\ \underline{[457]} \\
& - \ [257][354][317][237][351][537]\ \underline{[568]} \\
& - \ [257][354][237][351][358][735]\ \underline{[671]} \\
& + \ [257][354][317][356][351][735]\ \underline{[782]} \\
& + \ [257][354][317][356][257][357]\ \underline{[813]}
\end{aligned}
$$

$$
\begin{aligned}
& + ([157][327] - (1/2)[257][317])^2 \cdot [356][358][354] \\
& + (3/4)([257][317])^2 \cdot [356][358][354]
\end{aligned}
$$

Finally, we define

$$
g_{8_3} \quad := \quad \frac{4}{3} \cdot [356][358][354] \cdot f_{8_3} \quad \in \quad I_M^{\mathbf{Q}}.
$$

We multiply the second representation of f_{8_3} with $\frac{4}{3}[356][358][354]$ to obtain a representation of g_{8_3}.

From this representation we can see that g_{8_3} is a real final polynomial for M_{8_3}. For, observe that the first eight summands are contained in the ideal $I_M^{\mathbf{Q}}$, which is generated by the underlined brackets. The second to last summand is the product of $\frac{4}{3} > 0$ and a square, hence contained in $K[\Lambda(n,d)]^+$. Finally, the last summand equals

$$
([257][317] \cdot [356][358][354])^2 \quad \in \quad (S_M^{\mathbf{Q}})^2.
$$

We have

$$
g_{8_3} \quad \in \quad I_{8,3}^{\mathbf{Q}} \cap (I_M^{\mathbf{Q}} + \mathbf{Q}[\Lambda(8,3)]^+ + (S_M^{\mathbf{Q}})^2)
$$

and thus according to Theorem 4.19, the matroid M_{8_3} is not realizable over any ordered field.

4.3. Oriented matroids, the Steinitz problem and symmetric realizations

In this section we introduce final polynomials for (partial) oriented matroids, and we prove that a final polynomial exists in every nonrealizable case. With the results of Section 2.2 this yields a theorem of the alternative also for the Steinitz problem of deciding the polytopality of spheres. Furthermore, we formulate some remarks and an interesting open problem on symmetric realizations.

We proved the nonrealizability of the oriented matroid D_3^{10} in Section 3.3 by establishing a polynomial $f \in I_{10,3}^{\mathbf{Q}}$ such that in the expansion of f as the sum of monomials, all brackets have the same sign prescribed from D_3^{10}. The semi-algebraic Nullstellensatz of E. Becker (Theorem 4.11) tells us how to define final polynomials for the general case. As in the definition of real final polynomials in the previous section, there is no loss in generality in restricting ourselves to the ordered field $K = \mathbf{Q}$.

Given a partial oriented matroid

$$\chi \quad : \quad \Lambda(n,d) \quad \to \quad \{-1, 0, +1, *\},$$

let $H_\chi^{\mathbf{Q}}$ denote the multiplicative semi-group with unit generated in $\mathbf{Q}[\Lambda(n,d)]$ by the positive brackets

$$\{\ [\lambda]\ \mid\ \chi(\lambda) = +1\ \},$$

the negated negative brackets

$$\{\ -[\lambda]\ \mid\ \chi(\lambda) = -1\ \}$$

and the positive rationals $\mathbf{Q}^+$. Define $T_\chi^{\mathbf{Q}}$ to be the semi-ring in $\mathbf{Q}[\Lambda(n,d)]$ which is generated by $H_\chi^{\mathbf{Q}}$ and the squares $\mathbf{Q}[\Lambda(n,d)]^2$. Finally, write as before

$$I_\chi^{\mathbf{Q}} \quad := \quad < \{\ [\lambda] \in \Lambda(n,d)\ \mid\ \chi(\lambda) = 0\ \}\ >$$

for the ideal generated by all non-bases brackets.

A polynomial $f \in \mathbf{Q}[\Lambda(n,d)]$ is called a *final polynomial* for χ if

$$f \quad \in \quad I_{n,d}^{\mathbf{Q}} \cap (\ I_\chi^{\mathbf{Q}} + H_\chi^{\mathbf{Q}} + T_\chi^{\mathbf{Q}}\ ).$$

(Compare with the example in Section 3.3.) With this definition we have

Theorem 4.21. *The following statements are equivalent for a partial oriented matroid χ.*
(a) There exists a final polynomial for χ with coefficients in the rationals $\mathbf{Q}$.
(b) χ cannot be coordinatized over any ordered field.
(c) χ cannot be coordinatized over the real algebraic numbers $\mathbf{A} = \widetilde{\mathbf{Q}}$.

Proof. The equivalence of (b) and (c) follows either from Tarski's quantifier elimination [154] or from the results in [12] concerning the maximal real spectrum of an affine $\mathbf{Q}$-algebra.

The proof of the equivalence of (a) and (c) is based on Corollary 4.12, the alternate version of Becker's semi-algebraic Nullstellensatz. We write

$$\mathcal{R}^{\mathbf{A}}(\chi) \quad := \quad \{\xi \in G_{n,d}^{\mathbf{A}} \mid \operatorname{sign} [\lambda](\xi) = \chi(\lambda) \text{ whenever } \chi(\lambda) \neq *\}$$

for the *realization space* of χ with respect to $\mathbf{A}$. In applying the set-up in the end of Section 4.2 to this situation, we have $K = \mathbf{Q}$, $L = \mathbf{A}$, $V = G_{n,d}^{\mathbf{A}}$, $W = \mathcal{R}^{\mathbf{A}}(\chi)$,

$H = H_\chi^{\mathbf{Q}}$, $T = T_\chi^{\mathbf{Q}}$, and $I = I_\chi^{\mathbf{Q}}$, where each set in $\mathbf{Q}[\Lambda(n,d)]$ is identified with the corresponding set in the coordinate ring $\mathbf{Q}[G_{n,d}]$ of the Grassmann variety.

By Corollary 4.12, statement (c) in Theorem 4.21 is equivalent in $\mathbf{Q}[\Lambda(n,d)]$ to

$$(H_\chi^{\mathbf{Q}} + T_\chi^{\mathbf{Q}}) \cap I_\chi^{\mathbf{Q}} \neq \emptyset.$$

Since computations in $\mathbf{Q}[G_{n,d}]$ are carried out in $\mathbf{Q}[\Lambda(n,d)]$ modulo the ideal $I_{n,d}^{\mathbf{Q}}$, this is equivalent to the existence of a final polynomial $f \in \mathbf{Q}[\Lambda(n,d)]$. $\square$

With Proposition 2.6 we obtain the following immediate corollary for the Steinitz problem. A *final polynomial* for a matroid sphere S is a final polynomial for its associated partial oriented matroid χ_S.

Corollary 4.22. *Given a matroid sphere S, then one and only one of the following statements holds.*
(a) S is polytopal (over $\mathbf{A}$).
(b) There exists a final polynomial with rational coefficients for S.

Example 4.23. (Bokowski & Garms [24]) *Altshuler's sphere M_{425}^{10}.*

This sphere played a key role when the first final polynomials for more complicated geometric examples were searched, see also Section 8.1 for historical remarks about this sphere. Due to its high symmetry, the sphere M_{425}^{10} is of special interest among all neighborly 3-spheres with 10 vertices in the classification of A. Altshuler [3]. In [24] J.Bokowski & K.Garms obtained the following non-polytopality proof for M_{425}^{10}. Here we give a list of facets together with the intrinsic orientation of the sphere.

1230 −	1239 +	1260 +	1269 −	1345 −
1348 +	1350 −	1389 +	1450 +	1480 −
1670 +	1679 −	1790 +	1890 −	2345 +
2348 −	2357 +	2360 +	2367 −	2389 −
2458 −	2578 −	2678 +	2689 +	3560 +
3567 −	4560 −	4567 +	4579 +	4589 −
4670 −	4790 −	4890 +	5789 −	6789 +

Table 4-1. Oriented facets of Altshuler's sphere M_{425}^{10}.

Write $\chi : \Lambda(10,5) \rightarrow \{-1, +1, *\}$ for the associated partial oriented matroid. Consider the following polynomial

$$f := [16\overline{78}0][23\overline{46}0][24\overline{58}0][\overline{2}4689] \; - \; [12\overline{36}0][46\overline{78}0][24\overline{58}0][\overline{2}4689]$$
$$[23\overline{46}7][26\overline{89}0][24\overline{58}0][14\overline{68}0] \; - \; [\overline{2}6789][23\overline{46}0][24\overline{58}0][14\overline{68}0]$$
$$[23\overline{89}0][24\overline{56}8][\overline{2}4670][14\overline{68}0] \; - \; [23\overline{45}8][26\overline{89}0][\overline{2}4670][14\overline{68}0]$$
$$[45\overline{68}9][1\overline{2}480][\overline{2}4670][23\overline{68}0] \; - \; [14\overline{89}0][24\overline{56}8][\overline{2}4670][23\overline{68}0]$$
$$[1\overline{2}450][46\overline{78}0][\overline{2}4689][23\overline{68}0] \; - \; [45\overline{67}0][1\overline{2}480][\overline{2}4689][23\overline{68}0]$$

The signs with respect to χ of all brackets occurring in f can be read off from the intrinsic orientation of M^{10}_{425}. For example, $\chi(16780) = -\chi(1670\overline{8}) = -1$ because the facet 1670 is positively oriented. Checking the brackets shows that that all monomials of f are positive with respect to χ, and hence $f \in H^{\mathbf{Q}}_{\chi} + T^{\mathbf{Q}}_{\chi}$.

By expanding the three-term syzygies $\{\ldots | \ldots\}$, we obtain the identity

$$
\begin{aligned}
[24680] \cdot f \; = \; & + \{680|1742\} \cdot [26034][24085][26894] \\
& + \{260|1384\} \cdot [68074][24085][26894] \\
& + \{246|3708\} \cdot [26809][24085][68041] \\
& + \{268|9704\} \cdot [26034][24085][68041] \\
& + \{280|9346\} \cdot [24856][46072][68041] \\
& - \{248|3506\} \cdot [26809][46072][68041] \\
& + \{468|9520\} \cdot [48012][46072][26038] \\
& + \{480|9162\} \cdot [24856][46072][26038] \\
& + \{240|1568\} \cdot [68074][26894][26038] \\
& - \{460|5782\} \cdot [48012][26894][26038]
\end{aligned}
$$

which shows that $[24680] \cdot f$ is contained in $I^{\mathbf{Q}}_{10,5}$.

The 5-tuple $[24680]$ is determined (compare Section 3.3) to be negative in χ by the three-term syzygy

$$g \; := \; [24680][2345\overline{6}] \; - \; [234\overline{6}8][\overline{2}4560] \; + \; [2456\overline{8}][234\overline{6}0] \quad \in \quad I^{\mathbf{Q}}_{10,5}.$$

This shows that the polynomial

$$h \; := \; (\, g - [24680][23456] \,) \cdot f$$

is contained in $I^{\mathbf{Q}}_{10,5}$ as well as in $H^{\mathbf{Q}}_{\chi} + T^{\mathbf{Q}}_{\chi}$. Hence h is a final polynomial for M^{10}_{425}, which proves, by Corollary 4.22, that this sphere is not polytopal. $\qquad \square$

Let us close this section by outlining a few algebraic aspects of symmetric realizations. By a result of P. Mani [110], every 2-sphere S can be realized as the boundary complex of a polytope $P \subset \mathbf{R}^3$ with the property that every combinatorial automorphism of S induces an affine automorphism of P. Bokowski, Ewald & Kleinschmidt [23] showed that this result does not generalize to higher dimensions. Their proof is based on Kleinschmidt's (polytopal!) 3-sphere S with 10 vertices such that, in our language, the symmetry group of the associated partial oriented matroid χ_S is properly contained in the symmetry group of S. This shows that their counterexample is of combinatorial nature, and in a sence it would be nice to have a "more geometric" counterexample as precisely described below. The sphere plays also an essential part in the isotopy result given in Chapter 6.

Let Σ_n denote the symmetric group on $\{1, 2, \ldots, n\}$. There is a canonical action of Σ_n on $\Lambda(n, d)$ given by

$$\sigma\,[\lambda_1 \lambda_2 \ldots \lambda_d] \quad := \quad [\,\sigma(\lambda_1)\,\sigma(\lambda_2) \ldots \sigma(\lambda_d)\,].$$

To any subgroup G of Σ_n, we assign the ideal

$$J_G \quad := \quad < \{\,[\lambda] - \sigma[\lambda] \mid \lambda \in \Lambda(n, d),\ \sigma \in G\,\} > .$$

in $\mathbf{Q}[\Lambda(n, d)]$. The associated variety $\mathcal{V}(J_G) \subset \wedge_d \mathbf{Q}^n$ is the *linear subspace* of all d-vectors whose Plücker coordinates are symmetric with respect to G.

Naturally, Σ_n acts also on the rank d oriented matroids on $E = \{1, 2, \ldots, n\}$ by $(\sigma\chi)(\lambda) := \chi(\sigma^{-1}(\lambda))$. The *symmetry group* G_χ of an oriented matroid χ is defined in the obvious way :

$$G_\chi \quad := \quad \{\,\sigma \in \Sigma_n \mid \sigma\chi = \chi \text{ or } \sigma\chi = -\chi\,\}.$$

For an interesting practical application the reader is refered to Bokowski & Eggert [22]. In that paper symmetric embeddings of triangulated tori are studied by means of oriented matroid symmetry.

If the oriented matroid χ is realizable, i.e. $\mathcal{R}^{\mathbf{A}}(\chi) \neq \emptyset$, and G is a subgroup of Σ_n, then the space $\mathcal{R}^{\mathbf{A}}(\chi) \cap \mathcal{V}(J_G)$ of symmetric realizations of χ with respect to G is empty if G is not a subgroup of G_χ. Does the converse hold ? Is it true that for every realizable oriented matroid χ and every symmetry π of χ there exists a realization of χ in which π corresponds to an affine automorphism ? Although no counterexample is known to the author so far, we conjecture that such oriented matroids exist.

Conjecture 4.24. *There exists an oriented matroid χ and a group $G \subset G_\chi$ of symmetries of χ such that*

$$\mathcal{R}^{\mathbf{A}}(\chi) \neq \emptyset \quad but \quad \mathcal{R}^{\mathbf{A}}(\chi) \cap \mathcal{V}(I_G) = \emptyset.$$

4.4. On the Grassmann stratification and a problem of N. White

This section further explores the algebraic geometry of matroids. We discuss some questions related to N.White's paper [163] on the transcendence degree of matroid coordinatizations and the work of Gelfand et.al. [68] on the Grassmann stratification of $G_{n,d}^{\mathbf{C}}$. In particular, we solve a problem of N.White [142, Section 3] by constructing a counterexample to the "intuitive feeling that the algebraic object corresponding to a weak map is a specialization map", compare also [80, Commentary I.6] and [140, Exercise 9.2].

Recall that White's *bracket ring* of a matroid G with coefficients in some field K is our weak coordinate ring $K[G_{n,d}]/I_G^K$, and a prime ideal $p \subset B_G^K$ corresponds to a *proper coordinatization* iff $p \cap S_G^K = \emptyset$ for $p \supset I_G^K$ in $K[G_{n,d}]$.

Problem 4.25. (White, [142, Section 3]) *Given two matroids F and G, where $F \leq G$ in the weak-map ordering, and each can be coordinatized properly over some extension of K, then do there exist primes p_G and p_F of B_G^K properly coordinatizing G and F (resp.) with $p_G \subset p_F$?*

It is proved in [142, Theorem 4.3] that the answer is "yes" if F and G are binary, that is, if $K = \mathbf{GF_2}$. White gives the following motivation for this problem which might also be helpful for our discussion [142, Example 3.4].

Let G the rank 3 matroid on $E = \{1, 2, 3, 4, 5\}$ with non-bases [124] and [135], and let $x_1, \ldots, x_5$ be algebraically independent over K. Consider the realization matrix

$$M = \begin{pmatrix} 1 & 0 & 0 & x_2 & x_4 \\ 0 & 1 & 0 & x_3 & 0 \\ 0 & 0 & x_1 & 0 & x_5 \end{pmatrix}$$

for G over the field $K(x_1, \ldots, x_5)$. By now successively setting x_2, x_4, x_3 and x_5 equal to zero, we obtain a sequence of realizations of weak images of G, corresponding to an ascending chain of primes in B_G^K. Problem 4.25 now asks whether it is possible, given matroids $F \leq G$ (both realizable over some field extension of K), to find a (transcendental) realization of G which specializes to a realization of F.

Before describing our counterexample, we will prove a second reformulation of the problem which is, it seems, much closer to geometric intuition. This reformulation is related to the interesting work of Gelfand et.al. [68]. Let us outline a few basic ideas of their paper. For the remainder of this section we shall deal with the complex numbers $K = \mathbf{C}$ exclusively, with the understanding that most arguments generalize to arbitrary algebraically closed fields.

Two points ξ and η in the complex Grassmann manifold which realize the same matroid M are said to lie in the same *Grassmann stratum* Γ_M. So, the stratum Γ_M is the complex realization space $\mathrm{Max}(R_M^{\mathbf{C}})$ viewed as a subset (via Zariski homeomorphism) of $G_{n,d}^{\mathbf{C}}$. The coordinate ring $R_M^{\mathbf{C}}$ of the matroid M is then the coordinate ring of its Grassmann stratum Γ_M.

Note that, in general, Γ_M is *not* an (affine) *subvariety* of $G_{n,d}^{\mathbf{C}}$ because it is defined by equations *and* inequalities. Its vanishing ideal $\mathcal{I}(\Gamma_M)$ certainly contains the ideal $I_M^{\mathbf{C}}$, but, as we shall see below, in general these two ideals are not equal ! In other words, the Zariski closure $\overline{\Gamma_M}$ of Γ_M is not necessarily the weak realization space of M.

The set of strata

$$\Gamma_{n,d} \quad := \quad \left\{ \, \Gamma_M \mid M \text{ is a complex realizable rank } d \text{ matroid on } n \text{ elements} \, \right\}$$

which decomposes $G_{n,d}^{\mathbf{C}}$ into realization spaces of matroids will be refered to as *Grassmann stratification*. As the main results in [68], the authors derive two equivalent definitions of the Grassmann stratification, one of which describes $\Gamma_{n,d}$ as a multi-intersection of all possible Schubert cells with respect to the standard basis of $\mathbf{C}^n$.

In order to understand the topology of this stratification, it is important to study how the closure $\overline{\Gamma_G}$ of a stratum Γ_G intersects the other strata. In [53, Section 5.2] an example, based on harmonic quadrangles, is given of matroids G and F such that

$$\Gamma_F \cap \overline{\Gamma_G} \quad \neq \quad \emptyset \qquad \text{but} \qquad \Gamma_F \not\subset \overline{\Gamma_G}.$$

In other words, the Grassmann stratification is not a face-to-face cell decomposition of $G_{n,d}^{\mathbf{C}}$. (From now on we use F and G for matroids to make the connection to White's problem more suggestive.)

In any case, we have the following necessary condition for a matroid F whose stratum Γ_F intersects the closure of Γ_G.

Remark 4.26. *If $\Gamma_F \cap \overline{\Gamma_G} \neq \emptyset$, then $F \leq G$ in the weak map ordering of matroids.*

Proof. Given $\xi \in G_{n,d}^{\mathbf{C}}$, we write $m_\xi \in \mathrm{Max}(\mathbf{C}[G_{n,d}])$ for the corresponding maximal ideal in the coordinate ring of the Grassmann variety. Assume that $\xi \in \Gamma_F \cap \overline{\Gamma_G}$. The inclusion $\xi \in \Gamma_F$ implies $m_\xi \cap S_F^{\mathbf{C}} = \emptyset$, while $\xi \in \overline{\Gamma_G}$ implies $m_\xi \supset I_G^{\mathbf{C}}$. Hence $I_G^{\mathbf{C}} \cap S_F^{\mathbf{C}} = \emptyset$ and $[\lambda] \notin I_G^{\mathbf{C}}$ for every basis λ of F. Therefore, every basis of F is also a basis of G which proves the claim. $\qquad\square$

Now it is natural to ask whether the converse is true, that is given $F \leq G$, both complex coordinatizable, do Γ_F and $\overline{\Gamma_G}$ necessarily intersect ? This question turns out to be equivalent to White's Problem 4.25.

Lemma 4.27. *Let F and G be complex realizable rank d matroids on $E = \{1, 2, \ldots, n\}$. Then the following statements are equivalent.*
(a) *There exist primes p_G and p_F of $B_G^{\mathbf{C}}$ properly coordinatizing G and F (resp.) with $p_G \subset p_F$.*
(b) $\Gamma_F \cap \overline{\Gamma_G} \neq \emptyset$

Proof. It follows directly from the definition of the bracket ring $B_G^{\mathbf{C}}$ that statement (a) is equivalent to (a') There exist primes p_G and p_F of $\mathbb{C}[G_{n,d}]$ such that $S_F^C \cap p_F = \emptyset$, $S_G^C \cap p_G = \emptyset$, $I_F^C \subset p_F$, and $I_G^C \subset p_G \subset p_F$.

See Figure 4-2.

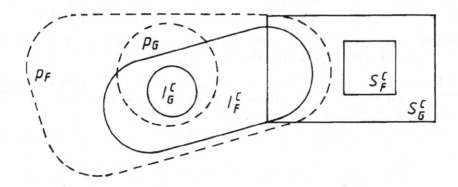

Figure 4-2. Condition (a') in visualized form.

Assume that (a') holds. According to a lemma of Krull [82, Lemma 4.4], we can assume that p_F is a maximal ideal. According to Hilbert's Nullstellensatz there exists a corresponding point $\xi \in G_{n,d}^C$, and it follows from the properties of p_F in (a') that $\xi \in \Gamma_F$.

Next observe the representation

$$\mathcal{I}(\Gamma_G) = \bigcap \{p \in \mathrm{Spec}(\mathbb{C}[G_{n,d}]) \mid p \cap S_G^C = \emptyset \text{ and } p \supset I_G^C\}. \quad (1)$$

This implies that $p_G \supset \mathcal{I}(\Gamma_G) = \mathcal{I}(\overline{\Gamma_G})$. Since p_G is a subset of p_F, this implies $p_F \supset \mathcal{I}(\overline{\Gamma_G})$, and therefore the point ξ which is associated with the maximal ideal p_F is contained in $\overline{\Gamma_G}$.

Conversely, assume that (b) holds. Pick $\xi \in \Gamma_F \cap \overline{\Gamma_G}$, and let $p_F \in \mathrm{Max}(\mathbb{C}[G_{n,d}])$ denote the corresponding maximal ideal. Then $p_F \supset I_F^C$ and $p_F \cap S_F^C \neq \emptyset$.

The point ξ being contained in $\overline{\Gamma_G}$, this is equivalent to $p_F \supset \mathcal{I}(\Gamma_G)$. Equation (1) implies that there exists a prime $p_G \subset p_F$ with $p_G \supset I_G^C$ and $p_G \cap S_G^C$.

This completes the proof. □

Let us express the result of Lemma 4.27 in more tangible terms. White's Problem 4.25 is equivalent to the following question :
   "Is there always a realization of F which can be approximated by realizations of G ?"
We just have to be a little bit careful about the topologies: If the answer was "yes" in the usual topology, then it would be "yes" also in the Zariski topology. But the answer being "no" in the usual topology does not automatically imply the same result for the Zariski topology. A careful ring-theoretic proof is required.

Theorem 4.28. *There exist complex realizable rank 3 matroids F and G on $E = \{1, 2, \ldots, 7\}$ with $F \leq G$ such that $\Gamma_F \cap \overline{\Gamma_G} = \emptyset$.*

Proof. Let G be the rank 3 matroid on $E := \{1, 2, \ldots, 7\}$ with non-bases [147], [257] and [367]. Let F be its weak image with non-bases [124] and $[ij7]$ for all $1 \leq i < j \leq 6$, that is 7 is a loop in F. See Figure 4-3.

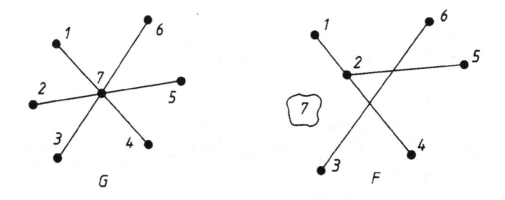

Figure 4-3. Two matroids F and G solving White's Problem 4.25

Consider the polynomial

$$
\begin{aligned}
&+ [125][623] \cdot \{[123][147] - [143][127] + [124][173]\} \\
&+ [124][125] \cdot \{[123][673] - [623][173] + [163][723]\} \\
&- [124][163] \cdot \{[123][527] - [523][127] + [125][723]\}
\end{aligned}
$$

which is zero in $\mathbf{C}[G_{7,3}]$. It is a linear combination of three-term syzygies $\{\ldots\}$ and thus contained in $I_{7,3}^{\mathbf{C}}$.

Expanding yields after cancellation of four summands the following representation of the zero polynomial in $\mathbf{C}[G_{7,3}]$.

$$+ [125][623][123]\,\underline{[147]}$$
$$+ [124][125][123]\,\underline{[673]}$$
$$- [124][163][123]\,\underline{[527]}$$
$$- [125][623][143]\,[127]$$
$$+ [124][163][523]\,[127] \quad = \quad 0$$

Since the underlined brackets generate the ideal $I_G^{\mathbf{C}}$, we conclude that $g \cdot [127] \in I_G^{\mathbf{C}}$ where

$$g \quad := \quad -[125][623][143] \; + \; \underline{[124]}\,[163][523].$$

According to the proof of Lemma 4.3, it is sufficient to show that condition (a') cannot hold in this situation. Assume on the contrary that there exist primes p_F and p_G as in (a'). Then we have $g \cdot [127] \in p_G$.

Since $[127] \in S_G^{\mathbf{C}}$, we have $[127] \notin p_G$. And, p_G being a prime ideal, this implies $g \in p_G \subset p_F$.

Since $[124]$ is not a basis of F, we have

$$\underline{[124]}\,[163][523] \quad \in \quad I_F^{\mathbf{C}} \quad \subset \quad p_F.$$

Subtracting g which is also contained in the ideal p_F, we obtain $[125][623][143] \in p_F$. On the other hand, this expression is a product of basis brackets with respect to F, and hence $[125][623][143] \in S_F^{\mathbf{C}}$. Thus we have $p_F \cap S_F^{\mathbf{C}} \neq \emptyset$, a contradiction. $\qquad\square$

Our counterexample in Theorem 4.28 might be somewhat unsatisfactory because the matroid F is not simple, that is F does not correspond to a geometric lattice. This shortcoming can be mended, however, by embedding both F and G in suitable rank 5 matroids.

Let us close this section by briefly discussing another very interesting problem concerning the topology of the Grassmann stratification which has been suggested by Gelfand et.al. in [53, Section 5.1].

Problem 4.29. *Given any complex realizable matroid M, is its stratum Γ_M necessarily a non-singular variety ? In other words, is the ring $R_M^{\mathbf{C}}$ regular ?*

Let us first note that the answer is "no" for the weak realization spaces $\mathrm{Max}(\mathbf{C}[G_{n,d}]/I_M^{\mathbf{C}})$. To see this, it is sufficient to show that for some matroid

M the bracket ring $B_M^{\mathbf{C}}$ is not regular. In [142, Section 5] White gives an example of a rank 3 matroid M on 9 elements such that $B_M^{\mathbf{C}}$ is not Cohen-Macaulay. This implies that $B_M^{\mathbf{C}}$ is not regular. Since both properties are local, and according to a well-known theorem in commutative algebra, every regular Noetherian local ring is Cohen-Macaulay [68, Theorem 8.2].

But we can obtain an even stronger result by a more direct argument. Since when we recall that according to the results of Section 2.1, every affine algebraic variety be can encoded as a Zariski dense subset in the realization space (modulo projective transformations) of some rank 3 matroid. With this method it is easy to construct rank 3 matroids M such that $\overline{\Gamma_M}$ has singularities.

However, in all the cases we have investigated so far, the singularities occurred *only* because of additional degeneracies, that is the singular points were always contained in $\overline{\Gamma_M} \setminus \Gamma_M$. In fact, in these cases the forming of the fraction $S_M^{\mathbf{C}}{}^{-1}$ corresponds to a "blowing-up" and yields a resolution of all singularities (cf. [68, Sect.I.4]). Nevertheless, we conjecture that there exist matroids M with singularities also in Γ_M.

It might be worth mentioning how the encoding procedure from Section 2.1 can be translated into the set-up of Gelfand et. al. [68]. The resulting algebraic universality result, Theorem 4.30, has been found independently from the results of the author by Mnëv [120].

The canonical action of the algebraic torus $H := (\mathbf{C}^*)^n$ on $G_{n,d}^{\mathbf{C}}$ corresponds to the action of the projective group on the vector configurations associated with the points on the Grassmann manifold, see [53, Sect.1.5]. The quotient Γ_M/H of a stratum Γ_M by this action is the projective realization space of M.

Theorem 4.30. *(Mnëv, Sturmfels) Given any affine algebraic $\mathbf{C}$-variety V defined over $\mathbf{Q}$, there exists an integer n and a rank 3 matroid M such that the projective realization space Γ_M/H of M is birationally isomorphic to V.*

Let us finally remark that if $K \in \mathbf{N}$ is a bound of the degree and the coefficients of integer polynomials defining V, then n is bounded by a polynomial in K.

Chapter V

GEOMETRIC METHODS

In this chapter we discuss geometric methods for the realization of oriented matroids and polytopes which are related to the study of the topology of the respective realization spaces. In the last section we consider an integral geometric approach to oriented matroids based on the Haar probability measure on the Grassmann manifold.

5.1. Geometric construction and topology of a class of oriented matroids

In this section we discuss topological properties and geometric constructions for a class of oriented matroids and partial oriented matroids. These are related to the famous *isotopy problem* which asks whether the realization space of an oriented matroid is necessarily path-connected [31], [72], [126].We refer to Chapter VI for a discussion of the general solution to the isotopy problem that has recently been found.

In the study of realization spaces of oriented matroids we are naturally led to the concept of *reducible points*. If an oriented matroid χ can be reduced to the simplex by deleting reducible points, then the realization space of χ is contractible and hence path-connected.

We give an easy algebraic proof for a result of Roudneff [111, Proposition 5.1], stating that points in sign-invariant pairs of rank 3 oriented matroids are always reducible. By inductive application we obtain the isotopy property and geometric realizations for all rank 3 oriented matroids with "sufficiently many" sign-invariant pairs. As induction hypotheses, it is proved that all rank 3 uniform oriented matroids with 7 points have contractible realization space, and we use Goodman & Pollack's result that all rank 3 oriented matroids with 8 points are realizable [69].

Let $\chi : E^r \to \{-1, 0, +1, *\}$ be a rank r partial oriented matroid on $E = \{1, 2, \ldots, n\}$. The *realization space* of χ, formally defined as

$$\mathcal{R}(\chi) \quad := \quad \{\, \Xi \in \wedge_r \mathbf{R}^n / \mathbf{R} \mid \Xi \text{ simple and sign } \Xi \text{ is above } \chi \,\},$$

is the topological space of all real realizations of χ modulo linear transformations. Thereby, the space $\wedge_r \mathbf{R}^n / \mathbf{R}$ of lines through the origin in the r-th exterior power of $\mathbf{R}^n$ can be identified with $(\binom{n}{r} - 1)$-dimensional real projective space. As an example note that the realization space of a simplex (χ as above with $n = r$) is just a point. Observe that $\mathcal{R}(\chi)$ is homeomorphic to $\mathcal{R}(\chi^*)$ under the canonical isomorphism between $\wedge_r \mathbf{R}^n$ and $\wedge_{n-r} \mathbf{R}^n$.

We say that χ is *reducible by* $e \in E$ provided every real realization of the minor $\chi \setminus e$ extends to a real realization to χ.

Proposition 5.1. *Let χ be as above and suppose that χ is reducible by $e \in E$. Then $\mathcal{R}(\chi)$ is homotopy equivalent to $\mathcal{R}(\chi \setminus e)$.*

Proof. Assuming that $E = \{1, 2, \ldots, n\}$, $e = n$, and that $\{1, 2, \ldots, r\}$ is a basis of χ, we can write the realization space of χ equivalently as

$$\mathcal{R}(\chi) \quad = \quad \left\{ (x_{r+1}, \ldots, x_n) \in \mathbf{R}^{(n-r)r} \mid (e_1, \ldots, e_r, x_{r+1}, \ldots, x_n) \text{ realizes } \chi \right\}$$

where $(e_1, e_2, \ldots, e_r)$ is any fixed basis of $\mathbf{R}^r$. Similarly and with the same basis $\mathcal{R}(\chi \setminus n)$ embeds into $\mathbf{R}^{(n-1-r)r}$. Every $(x_{r+1}, \ldots, x_{n-1}) \in \mathcal{R}(\chi \setminus n)$ defines a convex *extension cone*

$$\mathcal{C}(x_{r+1}, \ldots, x_{n-1}) \quad := $$
$$\left\{ \mathbf{x} \in \mathbf{R}^r \mid (e_1, \ldots, e_r, x_{r+1}, \ldots, x_{n-1}, \mathbf{x}) \text{ realizes } \chi \right\}$$

with respect to χ.

χ was assumed to be reducible with respect to n, and so $\mathcal{C}(x_{r+1}, \ldots, x_{n-1})$ is non-empty for all $(x_{r+1}, \ldots, x_{n-1}) \in \mathcal{R}(\chi \setminus n)$. Clearly, the extension cone depends continuously on its parameters, and so there exists a continuous mapping $f : \mathcal{R}(\chi \setminus n) \to \mathbf{R}^r$ such that $f(x_{r+1}, \ldots, x_{n-1}) \in \mathcal{C}(x_{r+1}, \ldots, x_{n-1})$ for all $(x_{r+1}, \ldots, x_{n-1}) \in \mathcal{R}(\chi \setminus n)$.

$\mathcal{R}(\chi \setminus n)$ being canonically homeomorphic to $\mathrm{Graph}(f) \subset \mathcal{R}(\chi)$, it is sufficient to prove that the "normalization map"

$$\pi : \mathcal{R}(\chi) \quad \to \quad \mathrm{Graph}(f) \quad \subset \quad \mathcal{R}(\chi)$$
$$(x_{r+1}, \ldots, x_{n-1}, x_n) \quad \mapsto \quad (x_{r+1}, \ldots, x_{n-1}, f(x_{r+1}, \ldots, x_{n-1}))$$

is homotopic to the identity on $\mathcal{R}(\chi)$. Since the extension cones are convex, the desired homotopy can be defined by the convex combination

$$\mathcal{R}(\chi) \times [0, 1] \quad \to \quad \mathcal{R}(\chi)$$
$$(x_{r+1}, \ldots, x_{n-1}, x_n; t) \quad \mapsto \quad (x_{r+1}, \ldots, x_{n-1}, t x_n + (1 - t) f(x_{r+1}, \ldots, x_{n-1})).$$

This proves Proposition 5.1. $\qquad\qquad\qquad\qquad\qquad\qquad\qquad\qquad\qquad\qquad\qquad$ $\square$

An ordering $(e_1, e_2, \ldots, e_n)$ of E is called a *reduction sequence* for χ if either $r = n$ or χ is reducible by e_n, and $(e_1, e_2, \ldots, e_{n-1})$ is a reduction sequence for $\chi \setminus e_n$. With an inductive application of Proposition 5.1 we obtain

Corollary 5.2. *If a partial oriented matroid χ admits a reduction sequence, then $\mathcal{R}(\chi)$ is contractible.*

Reduction sequences are very similar to the concept of *solvability sequences* as introduced in [31]. In fact, J. Richter showed that for rank 3 oriented matroids every solvability sequence induces a reduction sequence, and he uses this result to construct an oriented matroid without solvability sequence (but with the isotopy property) [28a].

The following example shows that the isotopy problem has an easy solution for partial oriented matroids, even in the uniform case.

Example 5.3. *There exists a rank 3 partial oriented matroid χ_5 with 5 points such that $\mathcal{R}(\chi_5)$ is disconnected.*

Define $\chi_5 : \Lambda(5,3) \rightarrow \{-1,+1,*\}$ by

[123] +	[124] +	[125] +	[134] *	[135] +
[145] +	[234] +	[235] *	[245] *	[345] +

Replacing all three "*" by "+" and by "−" respectively yields two realizable oriented matroids above χ_5. Their realization spaces cannot be connected by a path in the realization space of χ_5, see Figure 5-1.

□

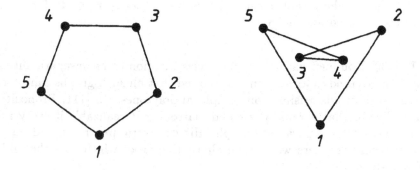

Figure 5-1. Two realizations of χ_5 contained in different path-components of $\mathcal{R}(\chi_5)$.

In the remainder of this section we are dealing exclusively with oriented matroids, not partial oriented matroids. Given an oriented matroid $\chi : E^r \to \{-1, 0, +1\}$ on E, two distinct elements e and e' of E are called *sign-invariant* in χ if they have always the same sign or always an opposite sign in all circuits of χ containing them. In the first case the sign-invariant pair $\{e, e'\}$ is called *covariant*, in the latter *contravariant*. The graph $\mathrm{IG}(\chi)$ on E whose edges are the sign-invariant pairs of χ is the *inseparability graph* of χ, see [17], [45]. Observe that a sign-invariant pair of χ is sign-invariant in every minor of χ and that $\mathrm{IG}(\chi) = \mathrm{IG}(\chi^*)$ where χ^* is the dual of χ.

Remark 5.4. *Every point in a rank 2 oriented matroid is reducible.*

Proof. Let $\chi : \Lambda(n, 2) \to \{-1, 0, +1\}$ be an oriented matroid. It is sufficient to show that n is reducible in χ. Let $x_1, x_2, \ldots, x_{n-1} \in \mathbf{R}^2$ be a realization of $\chi \setminus n$.
1st case: $\chi(i, n) = 0$ for all $i \in \{1, 2, \ldots, n - 1\}$. Define $x_n := 0$.
2nd case: There exist $i, j \in \{1, 2, \ldots, n-1\}$ such that $\chi(i, n) = 0$ and $\chi(j, n) \neq 0$. Define $x_n := \chi(j, n) \cdot \chi(j, i) \cdot x_i$.
3rd case: $\chi(i, n) \neq 0$ for all $i \in \{1, 2, \ldots, n - 1\}$. Then there exist $i, j \in \{1, 2, \ldots, n - 1\}$ such that $\{i, n\}$ and $\{j, n\}$ are sign-invariant pairs in χ respectively. After suitable reorientation we can assume w.l.o.g. that both pairs are contravariant. Define $x_n := x_i + x_j$.

It is easy to check that in each case $x_1, x_2, \ldots, x_{n-1}, x_n \in \mathbf{R}^2$ form a realization of the oriented matroid χ. □

In [130] Roudneff gave a complete classification of all inseparability graphs of uniform oriented matroids, and he showed that all such graphs can be realized in oriented matroids arising from simplicial polytopes. In [111, Proposition 5.1] he also showed that a rank 3 oriented matroid χ is realizable if every minor of χ has non-empty inseparability graph. His geometric proof is based on pseudo-line arrangements. Here we give an alternative proof which uses the underlying algebraic structure.

Proposition 5.5. *(Roudneff) Let χ be a rank 3 oriented matroid on E, and let $\{e, e'\} \subset E$ be a sign-invariant pair in χ. Then χ is reducible by e.*

Proof. Without loss of generality it may be assumed that $E = \{1, 2, \ldots, n\}, e' = n - 1, e = n$ and that $\{n - 1, n\}$ is a contravariant pair. We need to show that every linear realization $x_1, x_2, \ldots, x_{n-1} \in \mathbf{R}^3$ of $\chi \setminus n$ can be extended by a vector $x_n \in \mathbf{R}^3$ to a realization of χ.

Let $H \subset \mathbf{R}^3$ be the plane through 0 orthogonal to x_{n-1}, and let π denote the orthogonal projection onto H. Then $\pi(x_1), \pi(x_2), \ldots, \pi(x_{n-2}) \in H$ forms a linear realization of $(\chi/n - 1) \setminus n$. Let $y \in H$ be such that the oriented matroid of linear dependencies in H of $\{\pi(x_1), \pi(x_2), \ldots, \pi(x_{n-2}), y\}$ equals $\chi/(n - 1)$. Such a choice is possible according to Remark 5.4.

Now choose $\epsilon > 0$ such that $2\epsilon \cdot |\det(x_i, x_j, y)| \leq |\det(x_i, x_j, x_{n-1})|$ for all $i, j \leq n-2$, and define $x_n := x_{n-1} + \epsilon \cdot y$. First note that by the choice of $y = \pi(x_n)$, the determinant $\det(x_i, x_{n-1}, x_n) = \det(\pi(x_i), x_{n-1}, y)$ has the correct sign with respect to the chirotope of χ for all $i \leq n-2$. On the other hand, for all $i, j \leq n-2$,

$$\det(x_i, x_j, x_n) = \det(x_i, x_j, x_{n-1}) + \epsilon \cdot \det(x_i, x_j, y)$$

and hence, by the choice of ϵ

$$\text{sign} \det(x_i, x_j, x_n) = \text{sign} \det(x_i, x_j, x_{n-1}).$$

Since $\{n-1, n\}$ was assumed to be contravariant, this shows that all vector triples x_i, x_j, x_n have the correct orientation with respect to the chirotope of χ; this completes the proof of Proposition 5.5.

$\square$

Lemma 5.6.

(i) *Every rank 3 oriented matroid with $n \leq 8$ points is realizable.*

(ii) *Every uniform rank 3 oriented matroid with 7 points admits a reduction sequence.*

Proof. Part (i) follows from a result of Goodman & Pollack [69] and the well-known equivalence of arrangements of pseudo-lines and rank 3 oriented matroids, see Folkman & Lawrence [64].

Since the existence of a reduction sequence is a reorientation class invariant, it is sufficient in part (ii) to consider representatives of the 11 reorientation classes of uniform rank 3 oriented matroids with 7 points. Such representatives can be read off from Grünbaum's list [60, Figure 18.1.1] of simple arrangements with 7 lines in the projective plane as follows.

Consider the following 7×3-matrix with eight variable entries

$$\chi = \begin{pmatrix} 1 & 0 & 0 \\ 0 & 1 & 0 \\ 0 & 0 & 1 \\ x_4 & y_4 & 1 \\ x_5 & y_5 & 1 \\ x_6 & y_6 & 1 \\ x_7 & y_7 & 1 \end{pmatrix}.$$

For $i = 1, \ldots 11$, define χ_i^3 to be the rank 3 oriented matroid on $E = \{1, 2, \ldots, 7\}$ obtained from the matrix χ by replacing the variables as listed in row i of Table 5-1. Thereby, the x_j, y_j are ordered to yield a solvability sequence [31] with induced reduction sequence $(1, 2, 3, 7, 6, 5, 4)$. In addition we list the p-vectors [76] of the line arrangement associated with the χ_i^3. □

$$y_7: -1, y_6: -2, y_5: -3, y_4: -4, x_7: 1, x_6: 3, x_5: 6, x_4: 10, \quad p = (7, 14, 0, 0, 1)$$

$$y_4: 1, y_6: 2, y_7: 3, y_5: -1, x_7: -1, x_6: -2, x_5: 2, x_4: 1, \quad p = (11, 5, 5, 1)$$

$$y_4: 1, y_6: 2, y_7: 3, y_5: 4, x_7: -1, x_6: -2, x_5: -5, x_4: 8, \quad p = (9, 9, 3, 1)$$

$$y_7: 1, x_7: -1, x_6: -2, y_6: 3, y_4: 4, x_5: -3, y_5: 6, x_4: 1, \quad p = (8, 11, 2, 1)$$

$$y_7: -1, y_6: 1, x_7: 1, x_6: -2, x_5: -3, y_5: 4, y_4: -2, x_4: -4, \quad p = (7, 13, 1, 1)$$

$$y_6: 1, y_4: 2, y_7: 3, x_7: -1, x_6: -2, x_5: -3, y_5: 10, x_4: -\frac{1}{2}, \quad p = (10, 6, 6)$$

$$y_7: -1, y_6: -2, y_5: -\frac{1}{3}, x_7: -1, x_4: -2, x_6: -3, x_5: -\frac{2}{3}, y_4: 1, \quad p = (9, 8, 5)$$

$$y_7: 1, x_4: -1, x_7: -2, x_5: -3, x_6: -4, y_6: 3, y_5: 4, y_4: 7, \quad p = (8, 10, 4)$$

$$y_4: 1, y_7: 2, y_6: 4, y_5: 3, x_7: 1, x_6: 3, x_5: -1, x_4: 4, \quad p = (8, 10, 4)$$

$$y_7: -1, y_4: 1, y_6: 2, y_5: 3, x_7: -1, x_6: -2, x_5: 4, x_4: -9, \quad p = (7, 12, 3)$$

$$y_6: 1, y_7: 2, y_5: -1, x_7: -1, x_6: -2, x_5: -5, x_4: -6, y_4: 13, \quad p = (7, 12, 3)$$

Table 5-1. Solvability sequences for all uniform rank 3 oriented matroids with 7 points.

Using Lemma 5.6 as the induction hypothesis, we obtain from Lemma 5.2 and Proposition 5.5 the following result

Corollary 5.7. Let χ be a rank 3 oriented matroid matroid with n points.
(i) If χ has at least $n - 8$ sign-invariant pairs, then $\mathcal{R}(\chi) \neq \emptyset$, that is χ is realizable.
(ii) Moreover, if χ has at least $n - 7$ sign-invariant pairs, then the realization space $\mathcal{R}(\chi)$ of χ is contractible.

5.2. On neighborly polytopes with few vertices

Neighborly polytopes are of considerable importance in the theory of convex polytopes due to their connection with certain extremal properties, the most prominent of which is the Upper Bound Theorem, first established by P. McMullen [116]. While every neighborly $2k$-polytope with $2k + \beta$ vertices is combinatorially equivalent to a cyclic polytope if $\beta \leq 3$, I. Shemer [120, Theorem 6.1] proved that the number $c_n(2k + \beta, 2k)$ of combinatorial types of neighborly $2k$-polytopes with $2k + \beta$ vertices grows superexponentially as $\beta \to \infty$ ($n \geq 2$ fixed) and as $k \to \infty$ ($\beta \geq 4$ fixed). In fact, before the substantial improvements by Goodman & Pollack [71] and Alon [1], Shemer's "sewing" construction gave the best known lower bound for the number $c(n, d)$ of combinatorial types of *all* d-polytopes with n vertices.

A polytope P is said to have the *isotopy property* if for every polytope P' combinatorially equivalent to P, there exists a (continuous) path of polytopes of the same type from P to either P' or a mirror image of P'. It follows from Steinitz' analytic proof of his "Fundamentalsatz der konvexen Typen" [141] that all 3-polytopes have the isotopy property. Moreover, very recently S. Fischli obtained the result that the *unlabeled* realization space $[P]$ of a 3-polytope is connected if and only if P allows an orientation reversing automorphism. Otherwise $[P]$ has two components [111].

It is known that all cyclic polytopes, the prototypes of neighborly polytopes, do have the isotopy property [23, Example 5.2]. Here we apply the results of the previous section to prove a construction theorem and the isotopy property for a large class of neighborly $2k$-polytopes with $2k + 4$ vertices. The proof is based on the equivalence of universal edges in neighborly polytopes and sign-invariant pairs in the corresponding oriented matroids which will be established in Proposition 5.11.

Theorem 5.8. *Let S be a neighborly matroid $(2k-1)$-sphere with $2k+4$ vertices.*
(i) If S has at least $2k - 4$ universal edges, then S is polytopal.
(ii) If S has at least $2k - 3$ universal edges, then the polytopes with face lattice
    S fulfil the isotopy property.

So far, no neighborly $2k$-polytope with $2k + 4$ vertices is known to the author which violates the hypothesis of Theorem 5.8. Our results imply in particular an easy construction and the isotopy property for all 37 types of neighborly 6-polytopes with 10 vertices (Corollary 5.12). These polytopes have been classified by Bokowski & Shemer in [29].

Before we come to the discussion of neighborly polytopes let us make some remarks about polytopes in general. We have seen in Proposition 2.6 that for every polytopal sphere S there exists a partial oriented matroid χ_S such that the realizations of χ_S correspond to the polytopes with face lattice S. Therefore, we define the *convex realization space* of a sphere S as the realizations space $\mathcal{R}(\chi_S)$ of the associated partial oriented matroid. In particular, polytopes with face lattice S have the isotopy property if and only if $\mathcal{R}(\chi_S)$ is path-connected.

In [6] A.Altshuler and I.Shemer discuss constructions of polytopes based on *strongly coverable* and *P-coverable* point extensions. They remark in [4, Section 5] that all 1330 types of 4-polytopes with up to 8 vertices and all d-polytopes with $d + 2$ vertices can be obtained by these methods. On the other hand, it is easy to see that P-coverable and strongly coverable extensions for a sphere S correspond precisely to our notion of reducible points in the corresponding partial oriented matroid χ_S.

Consequently, we obtain the following result as a direct consequence of Corollary 5.2 and the results in [6].

Theorem 5.9. *All 4-polytopes with up to 8 vertices and all d-polytopes with $d + 2$ vertices have the isotopy property.*

A polytope $P \subset \mathbf{R}^d$ is called *m-neighborly* if the convex hull of every m-element set of vertices of P is a face of P. As is customary in the theory of neighborly polytopes, we restrict ourselves to the even-dimensional case [139], [149]. So, by a *neighborly* polytope we mean a k-neighborly $2k$-polytope $P \subset \mathbf{R}^{2k}$ for some k. Given a polytope P and a face ϕ of P, the quotient P/ϕ is a polytope whose face lattice is isomorphic to the upper segment $[F, P]$ in the face lattice of P. If P is a neighborly $2k$-polytope with set of vertices $V = \{v_1, v_2, \ldots, v_n\} \subset \mathbf{R}^{2k}$, an edge $\phi = \mathrm{conv}\{v_i, v_j\}$ is called *universal* if the quotient P/ϕ is a neighborly $2(k-1)$-polytope with $n - 2$ vertices.

These definitions obviously generalize to triangulated spheres and face lattices of oriented matroids. In the latter case, we shall use the standard term *contraction* rather than "quotient of matroid spheres", which is justified by the observation that quotients of polytopes correspond to contractions in the associated oriented matroids. Recall the following three easy facts on neighborly oriented matroids that are proved in [46] and [149].

Lemma 5.10.
(1) *An oriented matroid χ is m-neighborly if and only if $|C^+| \geq m + 1$ for all signed circuits C of χ.*
(2) *A rank $2k + 1$ oriented matroid χ is neighborly if and only if χ is uniform and $|C^+| = |C^-|$ for every signed circuit C of χ.*
(3) *Every neighborly rank $2k + 1$ oriented matroid with $2k + 3$ points is isomorphic to the alternating oriented matroid (or cyclic chirotope) which is associated with the cyclic polytope $C(2k + 3, 2k)$.*

Proposition 5.11. *Let χ be the neighborly oriented matroid associated with a neighborly $2k$-polytope $P = \mathrm{conv}\, V$. Then an edge $\phi = \mathrm{conv}\{v_i, v_j\}$ is a universal edge of P if and only if $\{v_i, v_j\}$ is a sign-invariant pair of χ.*

Proof.
If $|V| \leq 2k + 2$, then χ has either one or no signed circuit in which case the result is trivial. Hence, we may assume that $V = \{v_1, v_2, \ldots, v_n\}$ with $n \geq 2k + 3$.

Let $\{v_i, v_j\}$ be an sign-invariant pair of χ. First suppose that $\{v_i, v_j\}$ is a covariant pair. Pick a $2k + 3$-element subset V' of V such that $\{v_i, v_j\} \subset V'$, and let χ' the oriented matroid of affine dependencies on V'. Since every signed circuit of χ' is also a signed circuit of χ, $\{v_i, v_j\}$ is a covariant pair of χ' as well. χ' is the alternating rank $2k + 1$ oriented matroid on $2k + 3$ elements by Lemma 5.10 (3), and it follows from the results in [46] that χ' does not have covariant pairs. Hence, the sign-invariant pair $\{v_i, v_j\}$ is necessarily contravariant.

Every signed circuit $\widetilde{C}$ of $\chi/\{v_i, v_j\}$ can be written as $\widetilde{C} = C \setminus \{v_i, v_j\}$ where C is a signed circuit of χ. Since P is neighborly, we have $|C^+| = |C^-|$ by Lemma 5.10, and since $\{v_i, v_j\}$ is contravariant we can assume $v_i \in C^+$ and $v_j \in C^-$. Consequently, the sets $\widetilde{C}^+ = C^+ \setminus v_i$ and $\widetilde{C}^- = C^- \setminus v_j$ have the same cardinality, which implies with Lemma 5.10 (2) that P/ϕ is neighborly with $n - 2$ extreme points. Hence every sign-invariant pair of χ corresponds to a universal edge of P.

Conversely, let $\phi = \text{conv}\{v_i, v_j\}$ be a universal edge of P. Suppose there is a circuit C of χ such that $\{v_i, v_j\} \subset C^+$. Since P is neighborly, $|C^+| = |C^-| = k + 1$. According to the definition of universal edges, P/ϕ is neighborly, that is every circuit $\widetilde{C}$ of $\chi/\{v_i, v_j\}$ fulfils $|\widetilde{C}^+| = |\widetilde{C}^-| = k$. In picking the signed circuit $\widetilde{C} := C \setminus \{v_i, v_j\}$ of $\chi/\{v_i, v_j\}$, we have $|\widetilde{C}^+| = k - 1$, a contradiction. Hence, $\{v_i, v_j\}$ is a covariant pair of χ.

□

Proof of Theorem 5.8. Given a neighborly matroid $(2k - 1)$-sphere with $2k + 4$ vertices, there exists (by [128, Theorem 4.2] which generalizes a result of Shemer [120, Theorem 2.12]), a unique neighborly rank $2k + 1$ oriented matroid χ with $2k + 4$ vertices whose face lattice is S. In other words, the partial oriented matroid χ_S is below exactly one oriented matroid χ, and hence $\mathcal{R}(\chi_S) = \mathcal{R}(\chi)$.

S having at least $2k - 4$ universal edges implies with Proposition 5.11 that the dual χ^* is a uniform rank 3 oriented matroid with $2k + 4$ points and at least $2k - 4$ sign-invariant pairs. Hence χ and χ^* are realizable by Corollary 5.7 (i). Moreover, if S has at least $2k - 3$ universal edges, then Corollary 5.7 (ii) applies and the realization space $\mathcal{R}(\chi) = \mathcal{R}(\chi^*)$ is contractible. Hence S has the isotopy property.

□

In [29] J.Bokowski & I.Shemer gave a complete classification for all 37 (combinatorial types of) neighborly 6-polytopes with 10 vertices. Three proofs are given for the non-polytopality of 14 of the 51 listed neighborly 5-spheres with 10 vertices [21, Theorem 2]. Their third proof implies that the spheres $Q_{38}, Q_{39}, \ldots, Q_{51}$ are not matroid spheres because all 14 contain as a quotient the Brückner 3-sphere with 8 vertices which is not a matroid sphere [32]. The 37 matroid spheres $Q_1, Q_2, \ldots, Q_{37}$ all have at least three universal edges [21, Table 1], and so Theorem 5.8 yields a new polytopality proof and the isotopy property for these 37 spheres.

Corollary 5.12. *The oriented matroids with face lattices* $Q_1, Q_2, \ldots, Q_{37}$ *(see [29]) are realizable, and their realization spaces are contractible.*

We close this section with an interesting open problem on universal edges in neighborly polytopes. It can be shown that "universal edges" in neighborly spheres are a special case of the more general concept of "shrinkable edges" in general triangulated spheres, which has been of special interest in connection to the d-step conjecture. For a survey on this subject see Klee & Kleinschmidt [93].

The neighborly 4-polytope χ_{416}^{10}, see [32], with 10 vertices is the only known neighborly polytope without a universal edge, and it is also the smallest known example of a simplicial polytope without a shrinkable edge [77, Theorem 6.4]. It would be interesting to decide whether there exist simplicial d-polytopes without shrinkable edges with less than $d + 6$ vertices. In the context of this section we would like to suggest the following relaxation.

Problem 5.13. *Does there exist a neighborly $2k$-polytope P with $2k + 4$ vertices such that P has no universal edges ?*

5.3. A small counterexample to a conjecture of G.Ringel

In [101] M. Las Vergnas disproves a conjecture of G. Ringel by constructing a configuration of points in general position in the plane for which the x-coordinates cannot be "prescribed arbitrarily". Originally, this problem was posed in the polar version, that is whether the slopes can be prescribed for the lines in a simple arrangement $\mathcal{A}$ in the Euclidean plane [108, p.102].

To be more precise, let us consider point configurations $P_1, \ldots, P_n$ in general position in the real plane with x-coordinates $x_1 < \ldots < x_n$ such that

(R) for any sequence $x_1' < \ldots < x_n'$ of reals there are n points $P_1', \ldots, P_n'$ in general position in the plane with x-coordinates $x_1', \ldots, x_n'$ such that the oriented matroids of affine dependencies on $P_1, \ldots, P_n$ and $P_1', \ldots, P_n'$ are isomorphic.

Las Vergnas describes the geometric construction of a counterexample to (R) with $n = 32$ points. Moreover, he mentions the existence of a similar example with $n = 13$, and he asks whether $n = 13$ is minimal with this property. This question is particularly interesting for our discussion because if $n = 13$ were indeed minimal, then this would imply the isotopy property for uniform rank 3 oriented matroids with up to 13 points.

To see this, observe that prescribing the x-coordinates for a realizable rank 3 oriented matroid χ is equivalent to prescribing coordinates for the rank 2 minor Π/e of the point extension $\Pi := \chi \cup e$ of χ by a point e "at infinity in y-direction". Once the x-coordinates, that is coordinates for Π/e, are chosen, it is a linear programming feasibility problem to find y-coordinates, that is to extend the x_i to a coordinatization of Π.

When we dualize, then the property (R) for $\chi = \Pi/e$ (with e at infinity in y-direction) is equivalent to e being reducible in Π^*. If e is reducible in Π^* then, by Proposition 5.1, $\mathcal{R}(\Pi^*) = \mathcal{R}(\Pi)$ is homotopy equivalent to $\mathcal{R}(\Pi^* \setminus e) = \mathcal{R}(\Pi/e)$. The latter space is clearly contractible since Π/e is a rank 2 oriented matroid. This implies

Remark 5.14. *Let n_0 be the largest integer such that (R) is valid for all plane point configurations with up to n_0 points. Then the isotopy property holds for all rank 3 oriented matroids with up to $n_0 + 1$ points.*

However, it turns out that the answer to Las Vergnas' question is negative in the sense that n_0 is clearly less than 13.

Proposition 5.15. *There is a configuration with $n = 7$ points which does not satisfy condition (R)*

Proof. Let $P_1 = (-4, 12)$, $P_2 = (-3, 0)$, $P_3 = (-1, 7)$, $P_4 = (0, 11)$, $P_5 = (1, 7)$, $P_6 = (3, 0)$, $P_7 = (4, 12)$.

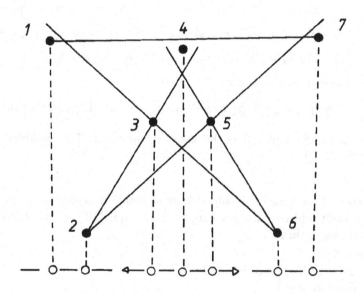

Figure 5-2. A point configuration violating Ringel's property (R)

Writing as before $[ijk]$ for the oriented area of the triangle $P_i P_j P_k$, we observe that

$$[135] > 0, \quad [163] > 0, \quad [147] > 0, \quad [234] > 0, \quad [275] > 0, \quad [456] > 0.$$

Choose $x_1' := -5$, $x_2' := -4$, $x_3' := -2$, $x_4' := 0$, $x_5' := 2$, $x_6' := 4$, $x_7' := 5$, and assume that there exist real numbers $y_1', \ldots, y_7'$ such that the oriented matroid of affine dependencies on $P_1' := (x_1', y_1'), \ldots, P_7' := (x_7', y_7')$ is isomorphic to χ. This implies that for all i, j, k the (oriented) area $[i'j'k']$ of the triangle $P_i' P_j' P_k'$ has the same sign as $[ijk]$. In particular, the following inequalities are satisfied:

$$[1'3'5'] = (x_3' - x_5') \cdot y_1' + (x_5' - x_1') \cdot y_3' + (x_1' - x_3') \cdot y_5' = -4 \cdot y_1' + 7 \cdot y_3' - 3 \cdot y_5' > 0$$

$$[1'6'3'] = (x_6' - x_3') \cdot y_1' + (x_3' - x_1') \cdot y_6' + (x_1' - x_6') \cdot y_3' = 6 \cdot y_1' + 3 \cdot y_6' - 9 \cdot y_3' > 0$$

$$[1'4'7'] = (x_4' - x_7') \cdot y_1' + (x_7' - x_1') \cdot y_4' + (x_1' - x_4') \cdot y_7' = -5 \cdot y_1' + 10 \cdot y_4' - 5 \cdot y_7' > 0$$

$$[2'3'4'] = (x_3' - x_4') \cdot y_2' + (x_4' - x_2') \cdot y_3' + (x_2' - x_3') \cdot y_4' = -2 \cdot y_2' + 4 \cdot y_3' - 2 \cdot y_4' > 0$$

$$[2'7'5'] = (x_7' - x_5') \cdot y_2' + (x_5' - x_2') \cdot y_7' + (x_2' - x_7') \cdot y_5' = 3 \cdot y_2' + 6 \cdot y_7' - 9 \cdot y_5' > 0$$

$$[4'5'6'] = (x_5' - x_6') \cdot y_4' + (x_6' - x_4') \cdot y_5' + (x_4' - x_5') \cdot y_6' = -2 \cdot y_4' + 4 \cdot y_5' - 2 \cdot y_6' > 0.$$

The positive linear combination

$$210 \cdot [1'3'5'] + 210 \cdot [1'6'3'] + 84 \cdot [1'4'7'] + 105 \cdot [2'3'4'] + 70 \cdot [2'7'5'] + 315 \cdot [4'5'6']$$

is zero which clearly contradicts the above inequalities. This completes the proof of Proposition 5.15. $\qquad\square$

In spite of this negative result, it is conceivable that for many realizable rank 3 oriented matroids coordinates can be found quickly by the following simple probabilistic algorithm.

Algorithm 5.16.
Input : A realizable rank 3 oriented matroid χ.
Output : Coordinates for χ.
  1. *Pick x-coordinates at random.*
  2. *Find either y-coordinates for a realization of χ or a dual solution to the LP-problem (as in the above proof of Proposition 5.15.) In the first case we are done; otherwise :*
  3. *Alter randomly those x-coordinates which were involved in the dual solution and go to 2.*

Naturally, it would be very nice to have some estimates for the *average case complexity* of Algorithm 5.16 or similar probabilistic coordinatization algorithms. In order to obtain such results, it is necessary to have a suitable measure for the "probability" of oriented matroids and point configurations. In the next and last section we shall develop a measure-theoretic approach to oriented matroids that seems to be appropriate for these questions.

5.4. An integral geometry approach to oriented matroids

In order to analyze the complexity of probabilistic procedures such as Algorithm 5.16, or to analyze the average case behaviour of any algorithm in computational geometry, one would like to be able to determine the probability that a configuration of n points in $\mathbf{R}^d$ has a certain property P. In particular, we are interested in assigning a value $0 \leq p(\chi) \leq 1$ to a (partial) oriented matroid χ such that with probability $p(\chi)$, n points chosen "at random" in $\mathbf{R}^d$ are a realization of χ. In this section we discuss some problems of classical models for questions of that type, and we define the probability $p(\chi)$ using the Haar measure $\mu_{n,d}$ on the Grassmann manifold $G^{\mathbf{R}}_{n,d}$. These ideas were suggested by J.E. Goodman and R. Pollack [private communication].

For an introduction to *integral geometry* and the study of geometrically defined measures see Santaló [133], Schneider [134]. Here we face the initial problem that there is, a priori and without any further requirements, no unique natural way to define such probability measures. In viewing point configurations as elements of $(\mathbf{R}^d)^n$, the first naive approach would be to consider the n-fold product $\lambda_d^n := \lambda_d \oplus \ldots \oplus \lambda_d$ of d-dimensional Lebesgue-measure to define the desired probabilities. This idea fails because for any oriented matroid χ, the subset of $(\mathbf{R}^d)^n$ corresponding to realizations has either λ_d^n-measure 0, or $+\infty$: λ_d^n is an infinite measure and thus does not induce a probability measure on $(\mathbf{R}^d)^n$.

There are several classical ways to circumvent this difficulty. One model which has extensively been studied in convexity (see Santaló [133]) assumes that the points x_i are independently equi-distributed within some compact convex subset K of $\mathbf{R}^d$. In other words, one considers the n-fold product $\lambda_{d,K}^n$ of the relative measure $\lambda_{d,K}(A) := \lambda_d(K \cap A)$ on $\mathbf{R}^d$. Many important results have been obtained in this setting, in particular for the case where K is the unit ball.

From our point of view this approach has two shortcomings. First, the measure is clearly not invariant under the choice of K. As an example, consider Sylvester's famous problem to determine the probability p_K that the convex hull of four points chosen at random in a plane convex body K is a quadrangle, see Buchta [41]. It is not difficult to see that $p_K = 1 - 4 \cdot F_3(K)$ where $F_3(K)$ denotes the expectation value of the area of the convex hull of three random points. Blaschke showed that $\frac{35}{48\pi^2} \leq F_3(K) \leq \frac{1}{12}$, both estimates being sharp. These results indicate that p_K truly measures a property of the convex body K and *not* of the corresponding polytope or oriented matroid. Another problem is that the measure $\lambda_{d,K}^n$ on $(\mathbf{R}^d)^n$ is not invariant under the canonical action of the special linear group, compare Section 1.2.

One important application of integral geometry has been estimating the average case complexity of the simplex method for linear programming. The work of K.-H. Borgwardt [36], who showed that the simplex algorithm is polynomial in average, deals with random vectors in $\mathbf{R}^d$ with respect to the n-fold product of some rotation invariant probability distribution. In particular, he studies the d-dimensional normal distribution. Again, these measures do not seem suitable for defining the (affine) probability of oriented matroids because the origin is singled out as a special point.

We have seen earlier that vector configurations (modulo the linear group) correspond to points in the Grassmann manifold $G_{n,d}^{\mathbf{R}}$ of d-dimensional vector subspaces of $\mathbf{R}^n$. Let us see that this manifold has an additional structure which naturally yields the desired probability measure.

The action of the topological group $SO(n, R)$ on the topological space $G_{n,d}^{\mathbf{R}}$ is transitive, continuous and open. This turns $G_{n,d}^{\mathbf{R}}$ into a *homogenous* $SO(n, R)$-space isomorphic to the quotient of the group $SO(n, R)$ by the stabilizer of that action. For details see Schneider [134].

In other words, $G_{n,d}^{K}$ is homeomorphic to

$$SO(n, \mathbf{R}) / \left(SO(d, \mathbf{R}) \times SO(n - d, \mathbf{R}) \right),$$

and this homeomorphism induces a natural group structure on the Grassmann manifold. Being a quotient of the compact group $SO(n, \mathbf{R})$, the group $G_{n,d}^{K}$ is compact.

Recall the result from functional analysis that on every such group G there is a unique (up to scalar multiples) motion invariant finite Borel-measure μ_G called the *Haar measure* on G, see Rudin [114, Theorem 5.14], Schneider [134], Nachbin [122].

Theorem 5.17. *On every compact group G there exists a unique regular Borel probability measure μ which is left-invariant, right-invariant and inversion-invariant, i.e. μ satisfies the relations*

$$\int_G f(x) \, d\mu(x) \;=\; \int_G f(x \cdot g) \, d\mu(x) \;=\; \int_G f(g \cdot x) \, d\mu(x) \;=\; \int_G f(x^{-1}) \, d\mu(x)$$

for all $g \in G$ and every continuous real-valued function f on G.

Hence, there exists a unique regular Borel probability measure $\mu_{n,d}$ on the Grassmann manifold $G_{n,d}^{\mathbf{R}}$ which is invariant under the action of $SO(n, \mathbf{R})$ on $G_{n,d}^{\mathbf{R}}$.

Definition 5.18. *Given any rank d partial oriented matroid χ on $E = \{1, 2, \ldots, n\}$, we define the probability of χ as*

$$p(\chi) \;:=\; \mu_{n,d}(\mathcal{R}(\chi))$$

where the realization space $\mathcal{R}(\chi)$ is considered as Borel subset of $G_{n,d}^{\mathbf{R}}$.

It immediately follows that an oriented matroid χ has non-zero probability if and only if χ is uniform. If $d = 1$ or $d = n - 1$, the Grassmann manifold is just projective space with the induced Haar measure from the sphere, and so in this case the probability of oriented matroids can be calculated (in principle) by elementary volume computations.

In every non-trivial instance, however, it seems to be very difficult to say something about this non-constructively defined measure $\mu_{n,d}$, and we suggest further research in this direction.

Problem 5.19.

a) *Compute or estimate the probability of some rank d oriented matroids with n points where $1 < d < n - 1$.*

b) *Find an algorithm to generate "random" points (with respect to the Haar measure $\mu_{n,d}$) on the Grassmann manifold $G_{n,d}^K$ using a random number generator for the unit interval.*

Geometric intuition suggests the following conjecture which states that in all dimensions the oriented matroids arising from cyclic polytopes have the highest probablity.

Conjecture 5.20. *The maximum among the probabilities of all rank d oriented matroids with n points is attained by the probability $p(\chi^{n,d})$ of the alternating oriented matroid $\chi^{n,d} : \Lambda(n,d) \rightarrow \{+1\}$.*

V. Klee pointed out D. Gale's related and more geometric conjecture. It states that "the likelihood of getting neighborly polytopes increases rapidly with the dimension of the space" [52, p.262].

Chapter VI

RECENT TOPOLOGICAL RESULTS

In this chapter we discuss the recent solutions to the isotopy problems for oriented matroids, uniform oriented matroids, polytopes and simplicial polytopes. These problems have independently been studied and solved by a group of geometers in the West and a group of topologists in the U.S.S.R.

The discrete geometry community in Western Europe and Northern America did not know about the striking papers of Vershik [157] and Mněv [121] until the summer of 1988. Up to that date the authors were aware of the following main developments with respect to the isotopy problem.

The isotopy conjecture for line arrangements in the projective plane first appeared in a 1956 paper of G. Ringel [126]. It was rephrased for order types or rank 3 oriented matroids by J.E. Goodman & R. Pollack in 1984 [72], and the first affirmative results appeared in Bokowski & Sturmfels [31]. In the spring of 1987 J. Richter proved that every uniform rank 3 oriented matroid with $n \leq 8$ points has the isotopy property [124], and he extended this result to $n = 9$ in the realizable case, [125].

In the summer of 1987 N. White constructed a rank 3 oriented matroid on $n = 42$ points without the isotopy property [164]. In using the Lawrence construction (Section 2.2), this implied the existence of convex polytopes without the isotopy property. At that time the cases of uniform oriented matroids and simplicial polytopes remained open. During the I.M.A. workshop on "Computational Geometry", Minneapolis, September 1987, L. Lovász mentioned that a group in Leningrad was studying similar problems, and we wrote to A.M. Vershik for information on their work.

In December 1987 B. Jaggi & P. Mani-Levitska found an example of a uniform oriented matroid without the isotopy property with $n = 17$ [90]. An easy alternative proof was given in the spring of 1988 by N. White and B. Sturmfels [89], see Section 6.1. At that time B. Sturmfels also found the construction method for simplicial polytopes to be discussed in Section 6.2.

In June 1988 N. White received a short note from A.M. Vershik refering to the articles [121], [153], and [157], which appeared in the Springer Lecture Notes in Mathematics series in the fall of 1988. We learned that beyond solving the isotopy problem, N.E. Mněv had proved a very general and striking universality theorem about realization spaces (Theorems 6.9 and 6.10). Although not all theorems of [121] are proved in that article, it is clear that Mněv's crucial techniques are almost identical to several of the methods developed in the present exposition. In Section 6.3 we state the main results of [121], and we sketch their proofs, using the language of oriented matroids.

6.1. Oriented matroids without the isotopy property

This section gives complete proofs for the existence of oriented matroids and uniform oriented matroids without the isotopy property. It is based on the joint note of the second author with Jaggi, Mani-Levitska and White [89].

Let $\mathcal{R}(M)$ denote the space of all vector realizations $(x_1,\ldots,x_n) \in (\mathbf{R}^3)^n$ (modulo the action of $GL(\mathbf{R},3)$) of a rank 3 oriented matroid M on n points. As in the proof of Proposition 5.1, we may assume that $(1,2,3)$ is a positively signed basis of M. Then $\mathcal{R}(M)$ equals the space of $3 \times n$-matrices whose maximal minors have signs given by the chirotope of M and whose first three columns contain the 3×3-unit matrix. If M is uniform, then $\mathcal{R}(M)$ is an open subset of $\mathbf{R}^{3(n-3)}$. The *isotopy problem* for a realizable M asks whether its realization space $\mathcal{R}(M)$ is a connected topological space.

A rank 3 oriented matroid M on $E = \{x_1, x_2, \ldots, x_n\}$ is said to be *constructible* if (x_1, x_2, x_3, x_4) is a projective basis and the point x_t is incident to at most two lines spanned by $\{x_1, x_2, \ldots, x_{t-1}\}$ for $t = 5, 6, \ldots, n$.

Lemma 6.1. *There exists a constructible rank 3 oriented matroid M_{17} on 17 points such that $\mathcal{R}(M_{17})$ has two connected components.*

Proof. Let M_{17} be the oriented matroid of the matrix

$$A(t) := \begin{pmatrix} 1 & 0 & 0 & 1 & 0 & 1 & 1 & 0 & 1 & 1 & 1 & 1 & 5 & 5 & 0 & 1 & -1 \\ 0 & 1 & 0 & 1 & 1 & 0 & -1 & 2 & 2 & 2 & 4 & 6 & 0 & -1 & t & -t & t \\ 0 & 0 & 1 & 1 & 1 & 1 & 0 & 1 & 2 & 0 & 1 & 0 & 6 & 5 & 1 & 0 & t-1 \end{pmatrix}$$

where $\frac{1}{5} < t < \frac{1}{2}(1 - \frac{1}{\sqrt{5}})$ or $\frac{1}{2}(1 + \frac{1}{\sqrt{5}}) < t < \frac{4}{5}$. M_{17} is well-defined because in evaluating all 3×3-minors, we find that $A(t)$ has the same oriented matroid for all choices of t in the above specified range. Moreover, every other choice of t yields a different oriented matroid.

It is easy to verify that M_{17} is constructible and that M_{17} is projectively unique up to the one parameter t. This proves that $\mathcal{R}(M_{17})$ is homotopy equivalent to the disjoint union of two open intervals on the real line. $\square$

Next we will construct a uniform oriented matroid without the isotopy property. Lemma 6.1 is sufficient to prove the following:

Theorem 6.2. (Sturmfels, White) *Let M be a constructible rank 3 oriented matroid on n points. Then there exists a <u>uniform</u> rank 3 oriented matroid $\widetilde{M}$ on at most $4(n-3)$ points and a continuous surjective map $\mathcal{R}(\widetilde{M}) \to \mathcal{R}(M)$. Hence $\mathcal{R}(\widetilde{M})$ is disconnected whenever $\mathcal{R}(M)$ is disconnected.*

Proof. We define a sequence $M =: M_n, M_{n-1}, M_{n-2}, \ldots, M_5, M_4 =: \widetilde{M}$ of oriented matroids and maps between their realization spaces. Let $n \geq t \geq 5$. Then M_{t-1} is constructed from M_t as follows. First suppose that x_t is incident to exactly two lines $x_i \vee x_j$ and $x_k \vee x_l$ with $1 \leq i,j,k,l < t$. Using the notation of Billera & Munson [16], we let M'_t be the oriented matroid obtained from M_t by the four successive *principal extensions*

$$x_{t,1} := [x_t^+, x_i^+, x_k^+], \quad x_{t,2} := [x_t^+, x_i^+, x_k^-],$$
$$x_{t,3} := [x_t^+, x_i^-, x_k^-], \quad x_{t,4} := [x_t^+, x_i^-, x_k^+]. \tag{1}$$

These extensions can be carried out for <u>every</u> vector realization of M_t by setting $x_{t,1} := x_t + \epsilon_1 x_i + \epsilon_2 x_k$, $x_{t,2} := x_t + \epsilon_3 x_i - \epsilon_4 x_k$, $x_{t,3} := x_t - \epsilon_5 x_i - \epsilon_6 x_k$, $x_{t,4} := x_t - \epsilon_7 x_i + \epsilon_8 x_k$ where $1 \gg \epsilon_1 \gg \epsilon_2 \gg \ldots \gg \epsilon_8 > 0$. This implies that the *deletion map* $\Pi : \mathcal{R}(M'_t) \to \mathcal{R}(M_t)$ is surjective. Geometrically speaking, in every affine realization of M'_t, the intersection point x_t is "caught" in the quadrangle $(x_{t,1}, x_{t,2}, x_{t,3}, x_{t,4})$. Define $M_{t-1} := M'_t \setminus x_t$ by deletion of that point, and let $\pi : \mathcal{R}(M'_t) \to \mathcal{R}(M_{t-1})$ denote the corresponding map.

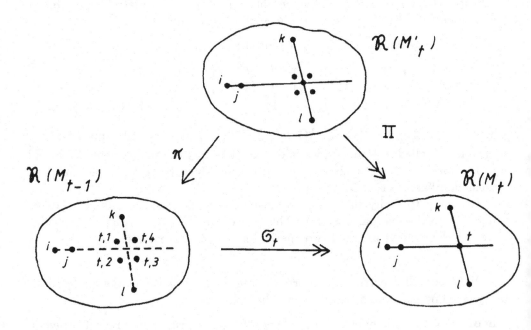

Figure 6-1. Illustration of the oriented matroids M_t, M'_t and M_{t-1}.

Next consider an arbitrary realization

$$X \quad := \quad (x_1, \ldots, x_{t-1}, \; x_{t,1}, x_{t,2}, x_{t,3}, x_{t,4}, \; x_{t+1,1}, \ldots, x_{n,4})$$

of M_{t-1}. As a consequence of the principal extension construction used in (1), $x_i \vee x_j$ and $x_k \vee x_l$ are the only lines spanned by $\{x_1, \ldots, x_{t-1}, x_{t+1,1}, \ldots, x_{n,4}\}$ which intersect the quadrangle $(x_{t,1}, x_{t,2}, x_{t,3}, x_{t,4})$. For any other such line the intersection point $x_t := (x_i \vee x_j) \wedge (x_k \vee x_l)$ is on the same side as $x_{t,1}, \ldots, x_{t,4}$. Therefore, $\sigma_t(X) := (x_1, \ldots, x_t, x_{t+1,1}, \ldots, x_{n,4}) \in \mathcal{R}(M_t)$.

Hence we have a well-defined continuous map

$$\sigma_t : \mathcal{R}(M_{t-1}) \to \mathcal{R}(M_t), \; X \mapsto \sigma_t(X).$$

Moreover, σ_t is surjective because $\Pi = \sigma_t \circ \pi$ is surjective.

It remains to define M_{t-1} and σ_t when x_t is incident to less than two lines in M_t. If x_t is on no such line, then we define $M_{t-1} := M_t$ and σ_t as the identity map. Finally, suppose that x_t is incident to only one line $x_i \vee x_j, 1 \le i, j < t$. In that case we replace (1) by setting $x_{t,1} := [x_t^+, x_k^+]$, $x_{t,2} := [x_t^+, x_i^+, x_k^-]$, $x_{t,3} := [x_t^+, x_i^-, x_k^-]$ for some $x_k \notin x_i \vee x_j$, and in the definition of the map σ_t we set $x_t := (x_i \vee x_j) \wedge (x_k \vee x_{t,1})$.

Iterating these constructions resolves all previous dependencies, and we obtain a uniform oriented matroid $\widetilde{M} := M_4$ on $4(n-3)$ or fewer points. Moreover, we have a continuous surjection $\sigma := \sigma_n \circ \sigma_{n-1} \circ \ldots \circ \sigma_5$ from $\mathcal{R}(\widetilde{M})$ onto $\mathcal{R}(M)$. □

In using a fairly straightforward procedure for doubling oriented matroids, one gets the following corollary whose proof will be omitted.

Corollary 6.3. *Given any integer C, there exists a uniform rank 3 oriented matroid $\widetilde{M}_C$ with $4(n-3)C$ points such that $\mathcal{R}(\widetilde{M}_C)$ has at least 2^C connected components.*

We close this section by stating the following result of P. Suvorov [153].

Theorem 6.4. *(Suvorov) There exists a rank 3 uniform oriented matroid M_{14} such that $\mathcal{R}(M_{14})$ is disconnected.*

Suvorov's proof of Theorem 6.4 uses the specific properties of a certain line arrangement, similarly to the construction of the 17-point uniform example by Jaggi & Mani-Levitska [90]. Naturally, the oriented matroids generated by general constructions such as Theorem 6.3 are larger than these specific examples. Theorem 6.4 together with the results of J. Richter [125] shows that the minimum number n_{min} of points in a rank 3 uniform oriented matroid without the isotopy property lies in the range $10 \le n_{min} \le 14$.

6.2. Simplicial polytopes without the isotopy property

The classical isotopy theorem due to E. Steinitz states that any two combinatorially equivalent 3-polytopes P and Q can be connected modulo reflections by a path of polytopes within the same class [141]. In Section 5.1 we proved this isotopy property for a large class of higher-dimensional polytopes. In using the results of the previous section and the Lawrence construction discussed in Chapter II, it is not difficult to construct non-simplicial polytopes without the isotopy property, see also [11].

In this section we describe a general construction technique for higher-dimensional <u>simplicial</u> polytopes without the isotopy property. Our exposition follows the proof given in the unpublished manuscript [151]. In the last part of this section, we prove that a specific simplicial 4-polytope with 10 vertices does not have the isotopy property. This result is due to N. Mněv [121], and J. Bokowski and A. Guedes de Oliveira [25] independently found the simple combinatorial proof to be presented here.

As before, a rank r oriented matroid χ on a set $E = \{1, \ldots, n\}$ of points will be identified with its chirotope map $\chi : E^r \to \{-, 0, +\}$, satisfying the signed bases exchange axiom given in Definition 1.12. Here we may assume that χ is realizable, that is there exist $x_1, \ldots, x_n \in \mathbf{R}^r$ such that $\chi(e_1, \ldots, e_r) = \det(x_{e_1}, \ldots, x_{e_r})$ for all $(e_1, \ldots, e_r) \in E^r$. The space of all such vector realizations of χ modulo the action of $GL(r, \mathbf{R})$ is denoted $\mathcal{R}(\chi)$ and called the *realization space* of χ. In using Plücker coordinates for the Grassmann manifold of r-flats in $\mathbf{R}^n$, $\mathcal{R}(\chi)$ embeds into a real projective space of dimension $\binom{n}{r} - 1$:

$$\mathcal{R}(\chi) \quad = \quad \{\, \xi \in (\wedge_r \mathbf{R}^n)/\mathbf{R} \mid \xi \text{ simple and } \operatorname{sign}(\xi) = \chi \,\}. \qquad (3)$$

Suppose that χ is *uniform* (i.e. $\operatorname{Im}(\chi) \subset \{-, +\}$) and *totally acyclic*, that is the positive hull $\mathcal{C}$ of $x_1, \ldots, x_n \in \mathbf{R}^r$ is a pointed cone. Then the face lattice of $\mathcal{C}$ is isomorphic to a simplicial $(r-2)$-sphere $\mathcal{S}$ on E which depends only on the oriented matroid χ. Conversely, given any simplicial $(r-2)$-sphere $\mathcal{S}$ on E, we define its *convex realization space* by

$$\mathcal{R}(\mathcal{S}) \quad := \quad \bigcup \{\, \mathcal{R}(\chi) \mid \mathcal{S} \text{ is the face lattice of } \chi \,\}, \qquad (4)$$

where the union is over all rank r matroid polytopes. This is equivalent to the definition given in Section 5.2, i.e. $\mathcal{R}(\mathcal{S})$ is the realization space of the unique (by Proposition 2.6) partial oriented matroid associated with $\mathcal{S}$. In particular, the space $\mathcal{R}(\mathcal{S})$ is disconnected if and only if Steinitz' isotopy theorem fails for the corresponding $(r-1)$-polytope $\mathcal{C} \cap H$ where $H \subset \mathbf{R}^r$ is a suitable affine hyperplane. We will prove the following theorem.

Theorem 6.5. (Sturmfels)
Let χ be a rank 3 oriented matroid on $E = \{1,\ldots,n\}$. Then there exists a simplicial sphere S_χ of dimension $n + \binom{n}{3} - 5$ with $n + \binom{n}{3}$ vertices such that if $\mathcal{R}(S_\chi)$ is connected then $\mathcal{R}(\chi)$ is connected.

In combining Theorem 6.5 with Suvorov's Theorem 6.4, we obtain the following

Corollary 6.6. *There exists a simplicial 373-sphere S with 378 vertices whose convex realization space $\mathcal{R}(S)$ is disconnected.*

Let us describe the construction of the sphere S_χ promised by Theorem 6.5. Proceeding by lexicographical order on the triples (i,j,k) where $1 \leq i < j < k \leq n$, we define *principal extensions* $e_{ijk} := [i^-, j^-, k^-]$. Let χ_{ext} denote the resulting __uniform__ rank 3 oriented matroid on $E' := E \cup \{e_{123}, e_{124}, \ldots, e_{n-2\,n-1\,n}\}$. (See [16], [31], [146] for combinatorial definitions of principal extensions of general oriented matroids.)

In the realizable case, principal extensions can be carried out geometrically as follows. Let $(x_1, \ldots, x_n) \in (\mathbf{R}^3)^n$ be any vector realization of χ. We define successively

$$x_{ijk} \; := \; -x_i - \epsilon \cdot x_j - \delta \cdot x_k \tag{5}$$

where $\delta \ll \epsilon \ll 1$ are sufficiently small positive numbers. Then

$$(x_1, \ldots, x_n, x_{123}, x_{124}, \ldots, x_{n-2\,n-1\,n}) \; \in \; (\mathbf{R}^3)^{n+\binom{n}{3}}$$

is a vector realization of χ_{ext}.

Let S_χ be the face lattice of the __uniform__ oriented matroid χ_{ext}^*, the dual of χ_{ext}. In other words, $(x_1, \ldots, x_n, x_{123}, x_{124}, \ldots, x_{n-2\,n-1\,n})$ is a Gale transform of the simplicial polytope with face lattice S_χ.

Lemma 6.7. *Let Ψ be any rank 3 oriented matroid on E' such that its dual Ψ^* has the face lattice S_χ. Then χ equals the restriction $\Psi|_E$ of Ψ to E.*

Proof. As in Section 2.2, we view the sphere S_χ as an oriented manifold with *intrinsic orientation* is_{S_χ}. is_{S_χ} is a map from the set of ordered facets of S_χ into $\{-1, +1\}$.

Let $1 \leq i < j < k \leq n$. By the definition of e_{ijk}, $\{i, j, k, e_{ijk}\}$ is a positive circuit of χ_{ext} and hence $F_{ijk} := E \setminus \{i, j, k, e_{ijk}\}$ is a facet of S_χ. We can write

$$\Psi(i,j,k) \; = \; \Psi^*(E \setminus \{i,j,k\}) \; = \; \pm \Psi^*(F_{ijk}, e_{ijk}) \; = \; \pm is_{S_\chi}(F_{ijk})$$

where the sign factor "$\pm$" does __not__ depend on the specific oriented matroid Ψ but only on (i, j, k). The last equation follows from our proof of Proposition 2.6.

In particular, we obtain $\Psi(i,j,k) = \chi_{ext}(i,j,k)$ and therefore $\Psi|_E(i,j,k) = \chi(i,j,k)$. $\qquad\square$

Proof of Theorem 6.4. Via the canonical *Hodge star* isomorphism $* : \wedge_r \mathbf{R}^n \to \wedge_{n-r} \mathbf{R}^n$ applied to formula (3), every oriented matroid has the same realization space as its dual, and we can write

$$\mathcal{R}(\mathcal{S}_\chi) \quad = \quad \bigcup \{\, \mathcal{R}(\Psi) \mid \mathcal{S}_\chi \text{ is the face lattice of } \Psi^* \,\}. \tag{6}$$

(In the language of polytope theory [76] the simplicial polytopes realizing $\mathcal{S}_\chi$ are parameterized by their *Gale transforms*).

We suppose that $\mathcal{R}(\mathcal{S}_\chi)$ is connected. Given any two vector realizations $(x_1^0 \ldots, x_n^0)$ and $(x_1^1, \ldots, x_n^1)$ of χ, we will show that they can be connected by a path in $\mathcal{R}(\chi)$. Using (5), we find vectors $x_{ijk}^0, x_{ijk}^1 \in \mathbf{R}^3$ such that
$X^0 := (x_1^0, \ldots, x_n^0, x_{123}^0, x_{124}^0, \ldots, x_{n-2\,n-1\,n}^0)$ and
$X^1 := (x_1^1, \ldots, x_n^1, x_{123}^1, x_{124}^1, \ldots, x_{n-2\,n-1\,n}^1)$ are realizations of χ_{ext}. X^0 and X^1 are in $\mathcal{R}(\mathcal{S}_\chi)$ by (6). Since $\mathcal{R}(\mathcal{S}_\chi)$ is connected, there exists a path

$$X : [0,1] \to \mathcal{R}(\mathcal{S}_\chi), \quad t \mapsto X^t = (x_1^t, \ldots, x_n^t, x_{123}^t, x_{124}^t, \ldots, x_{n-2\,n-1\,n}^t)$$

connecting X^0 and X^1. According to Lemma 6.8, we have $(x_1^t, \ldots, x_n^t) \in \mathcal{R}(\chi)$ for all $t \in [0,1]$. Hence $\mathcal{R}(\chi)$ is connected, and the proofs of Theorem 2 and Corollary 3 are complete. □

The number $\binom{n}{3} = O(n^3)$ in the statement of Theorem 6.5 may be replaced by the number $p_3(\chi)$ of triangular cells in the projective line arrangement associated with χ. To see this, we modify our above argument by defining χ_{ext} through successive principal extensions $e_{ijk} := [i^-, j^-, k^-]$ only for those triples which correspond to triangles in the line arrangement of χ.

Let $\mathcal{S}'$ be the face lattice of the resulting new χ_{ext}^*. According to the results of [131], the triangular cells of χ are in one-to-one correspondence with the mutations or local changes of χ.

Consider any realization $X_0 \in \mathcal{R}(\mathcal{S}')$ of χ_{ext}, and let X_1 be a configuration in the same connected component of $\mathcal{R}(\mathcal{S}')$. Under deletion of the new points e_{ijk}, both X_0 and X_1 are mapped into $\mathcal{R}(\chi)$, and moreover, they are mapped into the same connected component $\mathcal{R}(\chi)$. This proves that in Theorem 6.5 the number $\binom{n}{3}$ may be replaced by the number $p_3(\chi)$. According to results of Shannon and Roudneff (see [131]), we have the sharp bounds $n \leq p_3 \leq n(n-1)/3$ and hence $p_3(\chi) = O(n^2)$. We find that the simple arrangement of 17 lines given by of Jaggi and Mani-Levitska in [90] has 48 triangular cells. In using the above refined argument, we obtain a simplicial polytope without the isotopy property in dimension 61.

In the following we give a proof for the fact that this result can even be improved down to dimension 4.

Theorem 6.8. (N.E.Mněv / J.Bokowski, A.Guedes de Oliveira) *There is a simplicial 4-polytope P such that the realization space $\mathcal{R}(S)$ of its face lattice S is not path–connected.*

Remark 1.

According to Steinitz's theorem, this is the smallest dimension with this property. Mněv first mentioned this result in [121] but without providing a proof or a reference. Mněv's proof is currently planned to be included in a survey by Vershik and Mněv on "topology of configuration varieties and convex polytope varieties" in the Moskow journal "Functional Analysis and Applications". The proof we present here is due to J. Bokowski and A. Guedes de Oliveira, [25]. It is combinatorial in nature and uses only elementary arguments.

Remark 2.

The "disconnected" sphere S promised by Theorem 6.8 is the sphere which has been constructed by P. Kleinschmidt and which was the key-example in [23] and [4]. In these articles it was shown that in higher dimensions combinatorial automorphisms of polytopal spheres do exist which can not be derived as induced from any affine automorphism, compare also Chapter IV, after Example 4.23.

Remark 3.

The result can be carried over to higher dimensions. One has inductively to consider corresponding pyramids by starting with the 4-dimensional example.

Proof of Theorem 6.8.

The following expressions will be refered to as 3- or 4-term Grassmann-Plücker polynomials, respectively:

$$\{\tau \mid \lambda_1, \ldots, \lambda_4\} := [\tau, \lambda_1, \lambda_2][\tau, \lambda_3, \lambda_4] - [\tau, \lambda_1, \lambda_3][\tau, \lambda_2, \lambda_4] + [\tau, \lambda_1, \lambda_4][\tau, \lambda_2, \lambda_3].$$

$$\{\alpha \mid \beta \mid \gamma\} := [\alpha, \overline{\gamma_1}][\alpha, \beta, \gamma_1] - [\alpha, \overline{\gamma_2}][\alpha, \beta, \gamma_2] + [\alpha, \overline{\gamma_3}][\alpha, \beta, \gamma_3] - [\alpha, \overline{\gamma_4}][\alpha, \beta, \gamma_4].$$

Here we have used $\tau := \tau_1, \ldots, \tau_{d-2}$,
$\alpha := \alpha_1, \ldots, \alpha_{d-3}$, $\beta := \beta_1, \beta_2$, $\gamma := \gamma_1, \ldots, \gamma_4$, $\overline{\gamma_1} := \gamma_2, \gamma_3, \gamma_4$ $\overline{\gamma_2} := \gamma_1, \gamma_3, \gamma_4$ $\overline{\gamma_3} := \gamma_1, \gamma_2, \gamma_4$ $\overline{\gamma_4} := \gamma_1, \gamma_2, \gamma_3$.

Remember, the sign of a bracket $\chi[\lambda_1 \lambda_2 \cdots \lambda_d]$ within such a polynomial can be determined by the remaining signs of the other brackets within this polynomial if χ is required to be an oriented matroid.

Note that the change in the above Grassmann-Plücker polynomial, caused by a permutation of the points between two vertical lines, or between a vertical line and a bracket, does not change this property. The notation was chosen in order to avoid multiple checkings.

Next we provide the description of the sphere. We list all simplicial facets of its face lattice S below:

$\{1,2,3,4\}$ $\{1,2,3,7\}$ $\{1,2,4,8\}$ $\{1,2,6,7\}$ $\{1,2,6,8\}$ $\{1,3,4,7\}$ $\{1,4,5,6\}$
$\{1,4,5,8\}$ $\{1,4,6,7\}$ $\{1,5,6,8\}$ $\{2,3,4,8\}$ $\{2,3,7,0\}$ $\{2,3,8,9\}$ $\{2,3,9,0\}$
$\{2,6,7,9\}$ $\{2,6,8,9\}$ $\{2,7,9,0\}$ $\{3,4,5,7\}$ $\{3,4,5,8\}$ $\{3,5,7,0\}$ $\{3,5,8,0\}$
$\{3,8,9,0\}$ $\{4,5,6,7\}$ $\{5,6,7,9\}$ $\{5,6,8,0\}$ $\{5,6,9,0\}$ $\{5,7,9,0\}$ $\{6,8,9,0\}$

S allows (among others) a combinatorial automorphism ϕ, which is defined by the following permutation of the vertices: $(1,0)(2,5)(3,6)(4,9)(7,8)$.

The induced partial chirotope, which is determined by the convexity requirement of the sphere, does not determine a unique (complete) chirotope, in other words, the sphere is not rigid. But it is interesting to mention that even the uniquely determined partial chirotope violates the symmetry. Thus, the main result of [23] can already be established at this stage. This disproves also Conjecture 6.4 in [131].

The proof of our theorem proceeds as follows. We first list the essential orientations of the bases for our partial chirotope corresponding to the given boundary structure S. They are easily read from the following diagram, where some tetrahedral facets together with corresponding vertices are listed, similar to [5]. Compare also Section 2.2. We use any realization P of our combinatorially given polytope, see e.g. [23], and its corresponding image Q under the symmetry ϕ. In assuming that there is a continuous path connecting them, as the isotopy-property requires, we will get a contradiction.

For a facet $F = \{i_1, \ldots, i_4\}$, the sign $\chi[i_1, \cdots, i_4, j]$ is independent of the choice of $j \neq i_1, \ldots, i_4$.

The signs of these special brackets (sign[facet,point]) are all determined, up to reversing them all. W.l.o.g., we may assume that for the partial chirotope χ of P, we have $\chi[1,2,3,4,6] = -1$.

The signs we explicitly need, we can derive from the following scheme:

Here we use for the signs of the brackets $\chi[\lambda_1, \lambda_2, \cdots, \lambda_d]$ the shorter notation $\lambda_1 \lambda_2 \cdots \lambda_d$.

$$-1 = 1234\,6 = \underbrace{1234\,7 = -1237\,4}_{\text{by reordering}} = -1237\,5 = \underbrace{-1237\,6 = 1267\,3}_{\text{by reordering}} = \cdots \quad .$$

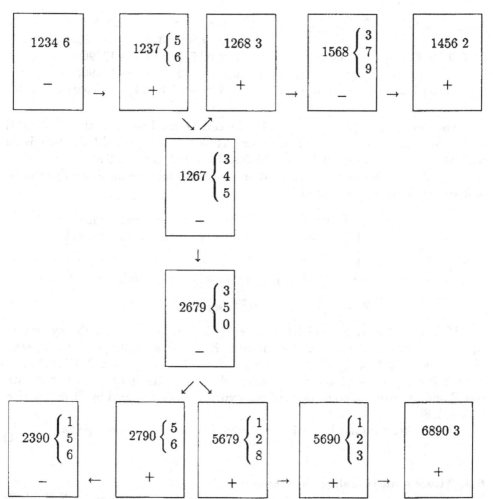

In the following we will show that the product of signs of a particular set of bases, namely $\chi[2,3,5,6,7] \times \chi[2,3,5,6,8]$, cannot be equal to 1. For that, we provide a list of appropriate Grassmann-Plücker-polynomials. For confirming our property, we use the above signs of the partial chirotope.

A first step implies $\chi[1,2,6,3,5] = -1$:

$$\{1,2,6 \mid 4,7,3,5\} =$$
$$\underbrace{[1,2,6,4,7]}_{+} [1,2,6,3,5] - \underbrace{[1,2,6,4,3]}_{+} \underbrace{[1,2,6,7,5]}_{-} + \underbrace{[1,2,6,4,5]}_{-} \underbrace{[1,2,6,7,3]}_{-} = 0$$

The polynomial $\{1,2,6 \mid 4,7,3,5\}$ determines 12635, written as

$$\{1,2,6 \mid 4,7,3,5\} \longrightarrow 12635.$$

1: $\chi[2,3,5,6,7] = \chi[2,3,5,6,8] = 1$

$\{2,6,9,\,|\,7,0,3,5\} \longrightarrow 23569;$

$\{5,6\,|\,2,3\,|\,7,1,8,9\} \longrightarrow$ contradiction.

2: $\chi[2,3,5,6,7] = \chi[2,3,5,6,8] = -1$

$\{2,5\,|\,9,0\,|\,1,3,6,7\} \longrightarrow 12590;$

$\{2,9,0\,|\,3,5,1,6\} \longrightarrow 12690;$

$\{6,9,0\,|\,2,5,1,3\} \longrightarrow 13690;$

$\{3,6\,|\,9,0\,|\,2,5,8,1\} \longrightarrow$ contradiction.

The polynomial $\{2,5 \mid 9,0 \mid 1,3,6,7\}$ can be used to show that $[2,3,5,6,7]$ cannot be equal to zero in any realization. Thus, we have proved that there is no realization of our sphere, such that $[2,3,5,6,7]$ is equal to $[2,3,5,6,8]$.

As before in Section 4 our realization space (modulo the linear group) can be written explicitly as a subset of $\mathbf{R}^{25}$

$$M : \;=\; \begin{pmatrix} 1 & 0 & 0 & x_{4,1} & 0 & 0 & x_{7,1} & x_{8,1} & x_{9,1} & x_{10,1} \\ 0 & 1 & 0 & x_{4,2} & 0 & 0 & x_{7,2} & x_{8,2} & x_{9,2} & x_{10,2} \\ \vdots & \vdots & & & & & & & \vdots & \vdots \\ 0 & 0 & 0 & x_{4,5} & 0 & 1 & x_{7,5} & x_{8,5} & x_{9,5} & x_{10,5} \end{pmatrix}$$

$$\mathcal{R}(S) \;=\; \{M \in \mathbf{R}^{25} : sign(...) = \chi(...)\}.$$

Pick any $P \in \mathcal{R}(S)$ and let $Q = \phi(P)$ be its image under the symmetry ϕ, $Q := (x_{\phi(1)}, \ldots, x_{\phi(10)})$. With indices P and Q we distinguish both cases. Notice that $[2,3,5,6,7]_Q = [2,3,5,6,8]_P$; so $\chi[2,3,5,6,7]_P = \chi[2,3,5,6,8]_Q = -\chi[2,3,5,6,8]_P = -\chi[2,3,5,6,7]_Q$, which shows that the images of our realizations P and Q lie on different sides of our hyperplane determined by $[2,3,5,6,7] = [2,3,5,6,8]$.

Hence, $\mathcal{R}(S)$ cannot be path–connected, which concludes our proof. □

6.3. Mnëv's universality theorem

In this section we discuss the striking universality theorem proved by N.E. Mnëv [121]. The statement says, roughly speaking, that arbitrary semi-algebraic $\mathbf{Q}$-varieties can be encoded in suitable oriented matroids or polytopes, and even in uniform oriented matroids or simplicial polytopes.

Two semi-algebraic $\mathbf{Q}$-varieties $V, W \subset \mathbf{R}^n$ are said to be *stably equivalent* if there is a piecewise biregular homeomorphism ϕ between V and $W \times \mathbf{R}^i$ for some $i \in \mathbf{N}$. The homeomorphism ϕ being locally biregular means that locally both ϕ and ϕ^{-1} can be defined by rational functions.

Clearly, stable equivalence implies homotopy equivalence. Following the arguments of Mnëv, we may replace "homotopy equivalent" by "stably equivalent" in all constructions described in Sections 5.1, 5.2, 6.1 and 6.2. In particular, "contractible" may be replaced by "homeomorphic to some $\mathbf{R}^i$ ".

The main results of [121] may be reformulated as follows.

Theorem 6.9. (Mnëv) *Let V be any semi-algebraic subvariety of $\mathbf{R}^n$, defined over $\mathbf{Q}$.*

(1) There exists a rank 3 oriented matroid χ such that V is stably equivalent to $\mathcal{R}(\chi)$.

(ii) For some $d \in \mathbf{N}$ there exists a $(d-1)$-sphere $\mathcal{S}$ with $d+4$ vertices such that V is stably equivalent to $\mathcal{R}(\mathcal{S})$.

This theorem can still be strengthened for oriented matroids and polytopes in general position.

Theorem 6.10. (Mnëv) *Let V be any semi-algebraic open subvariety of $\mathbf{R}^n$ defined by <u>sharp</u> polynomial inequalities over $\mathbf{Q}$.*

(1) There exists a <u>uniform</u> rank 3 oriented matroid χ such that V is stably equivalent to $\mathcal{R}(M)$.

(ii) For some $d \in \mathbf{N}$ there exists a <u>simplicial</u> $(d-1)$-sphere $\mathcal{S}$ with $d+4$ vertices such that V is stably equivalent to $\mathcal{R}(\mathcal{S})$.

In the following we summarize the proofs of the four universality theorems.

Statement (i) of Theorem 6.9 can be understood as an oriented analogue to our results in Section 2.1. There we gave the following explicit construction for encoding arbitrary algebraic $\mathbf{Q}$-varieties in suitable rank 3 matroids or plane projective configurations. Recall that we defined an alphabet L_n of polynomial formulas in n variables $x_1, x_2, \ldots, x_n$. This gave rise to a canonical surjection $\sigma : L_n \rightarrow \mathbf{Z}[x_1, x_2, \ldots, x_n]$.

In order to prove the substantially more difficult oriented version, Mnëv introduces a similar but more powerful set-up. His formal language $\widetilde{L}_n$ includes also the division operator, and hence he obtains a canonical surjection $\widetilde{\sigma} : \widetilde{L}_n \rightarrow \mathbf{Q}[x_1, x_2, \ldots, x_n]$.

Suppose that the semi-algebraic variety $V \subset \mathbf{R}^n$ is defined by k polynomials $f_i \in \mathbf{Q}[x_1, x_2, \ldots, x_n]$, $i = 1, 2, \ldots, k$. Using sophisticated regularity conditions and a very careful analysis, Mnëv then constructs a "suitable" k-tuple $(p_1, p_2, \ldots, p_k) \in (L_n)^k$ such that $\sigma(p_i) = f_i$. In applying the constructions explained in Section 2.1, he then finds a certain rank 3 matroid M which encodes $(p_1, p_2, \ldots, p_k)$.

The crux of the Mnëv's construction is the following: Among all oriented matroids with underlying matroid M, there is one oriented matroid χ such that the sign-pattern defining V from $(f_1, f_2, \ldots, f_k)$ is precisely reflected in χ. From this it follows that V and $\mathcal{R}(\chi)$ are stably equivalent.

The other three universality theorems can be derived as follows. Theorem 6.9 (ii) follows from Theorem 6.9 (i) using the Lawrence construction in Section 2.2, see also [11].

Mnëv's derivation of Theorem 6.10 (i) from Theorem 6.9 (i) in [103, Section 5] is almost identical to our construction in Section 6.1. It requires a little extra argument to see that the continuous surjection $\mathcal{R}(\widetilde{M}) \rightarrow \mathcal{R}(M)$ in Theorem 6.2 is indeed a stable equivalence. Finally, Theorem 6.10 (ii) can be derived from Theorem 6.9 (ii) using a construction as in Section 6.2.

Chapter VII

PREPROCESSING METHODS

In taking into account the very general setting of Computational Synthetic Geometry, there are still huge subclasses of problems for which specialized efficient algorithms are applicable. In this chapter we describe three successfully applied algorithms: the inductive generation of oriented matroids with special properties, the solvability sequence method leading to coordinates for oriented matroids, and a method of finding final polynomials by using linear programming techniques.

7.1 Generation of oriented matroids with specific properties

A geometric computational problem was called synthetic if the input consists of a combinatorial or geometric condition, and the output consists of either coordinates for such a geometric object or a proof that such an object is not realizable. This easily leads to the realizability problem of oriented matroids for a large class of problems. For instance, deciding an arbitrary (finite) lattice $\mathcal{L}$, whether $\mathcal{L}$ is the face lattice of a convex polytope with vertex coordinates in K, is algorithmically equivalent to deciding an arbitrary oriented matroid χ, whether χ is coordinatizable over K (Theorem 2.5.). In this particular case, there are straightforward algorithms (see 2.6.) for obtaining corresponding (partial) oriented matroids, and moreover, as experience shows, the CPU-time for this part can be neglected in compared to the time needed for the decision procedures which have to be applied afterwards.

On the other hand, there are cases in which determining a corresponding oriented matroid is a very hard problem in practice. Embedding problems for simplicial complexes as mentioned in Chapter 1 or more general, embedding problems for arbitrary geometric complexes are among these problems. For these cases an inductive generation of oriented matroids will be discussed in this section.

The main idea of this algorithmic construction is starting with a suitable minor of the oriented matroid we are looking for and appropriately extending it, afterwards. In the extreme case we start with just one basis, e.g. an oriented matroid with d points in rank d, and again in the extreme case we construct all possible extensions of the oriented matroid we started with, and we check whether we have found a feasible oriented matroid which in case of realizability would give us a solution to our problem. But very often the combinatorial or geometrical input passes on some properties to substructures which can often be used for reducing the class of minors to be considered in the algorithm. And sometimes we can choose w.l.o.g. a certain minor with more than d points to start with. Thus, the tree of all possible extensions can be cut substantially.

This general idea can be made more precise by looking at a special subclass of such problems and by considering the following Algorithm 7.2. We assume for simplicity that we already have an oriented matroid with $n - 1$-points admissible to our given combinatorial manifold $\mathcal{C}$ with n points. Before we make this more precise, and before presenting this algorithm, we introduce the notion of *hyperline configurations* in applying it to a particular example. Hyperline configurations are also closely related to Goodman and Pollack's notion "cluster of stars" [70, Problem 5.4].

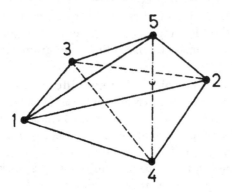

Figure 7-1. Double pyramid over the triangle 1,2,3.

Example 7.1. (Hyperline configuration) *We choose 5 points in $\mathbf{R}^3$ forming the vertices of a double pyramid over a triangle. Pick an ordered pair of vertices and project all 5 points along the ordered line determined by this pair of points onto a 2-dimensional affine plane orthogonal to this line. This determines for each line a cyclic order under which the projected vertices (not lying on this line) are seen from the line. It suffices to write only half of this circular sequence. This is done in the following for all oriented lines. We call it a complete list of hyperline configurations.*

$$(1,2 \mid 3,4,\bar{5}) \quad (1,3 \mid 2,5,\bar{4}) \quad (1,4 \mid 2,5,3) \quad (1,5 \mid 2,\bar{3},\bar{4}) \quad (2,3 \mid 1,4,\bar{5})$$

$$(2,4 \mid 1,\bar{3},\bar{5}) \quad (2,5 \mid 1,4,3) \quad (3,4 \mid 1,5,2) \quad (3,5 \mid 1,\bar{2},\bar{4}) \quad (4,5 \mid 1,\bar{2},3)$$

In particular, all orientations of simplices can be read off from this *complete list of hyperline configurations* in a straightforward manner. This abstract notion still makes sense for oriented matroids, and moreover, it is even a possible approach in defining chirotopes, since the 3-term Grassmann Plücker relations are encoded in this structure.

We consider only the uniform case in rank d. A *complete set of hyperline configurations* can be seen as a list of $\binom{n}{d-2}$ permutations of $E = \{1, 2, \ldots, n\}$ beginning with pairwise disjoint ordered $d - 2$-tuples together with an ordering and an assignment of signs for the remaining $n - d + 2$ elements of the permutation, written in the form

$$(\lambda_{l_1}, \ldots, \lambda_{l_{d-2}} \mid \mu_{l_1}, \ldots, \mu_{l_{n-d+2}})$$

with

$$\lambda_{l_1} < \ldots < \lambda_{l_{d-2}}$$

and using $\overline{\mu_{l_i}}$ in case of negative signs. Of course, we also think of extending the right part to a cycling ordering where half of this cycle is given and opposite numbers are equal except for the signs. It is easy to see that the following proposition holds.

Proposition 7.2. *An alternating map* $\chi : E^n \to \{-1, 0, 1\}$ *is a (uniform) oriented matroid if and only if there exists a complete set of hyperline configurations such that*

$$\chi[\lambda_{l_1}, \ldots, \lambda_{l_{d-2}}, \mu_{l_s}, \mu_{l_t}] = 1,$$

for all l and for all s, t with s < t.

The next example of an oriented matroid is written in terms of hyperline configurations to show and underline the compact form in which the data of an oriented matroid is stored this way. The example describes a matroid manifold of genus 4 with minimal number of vertices. A *matroid manifold* is an oriented matroid which in case of realizablity would yield a geometric (flat) manifold. Thus in our case we would get a geometric 2-manifold with minimal number of vertices, see [21] as well as Chapter VIII for more details.

It took a lot of CPU time to find this oriented matroid, but it was an essential step on the way to answering an open embedding problem. In the following we deal with the question of how to find such examples.

$(1, 2|3, 4, 8, 0, 9, \overline{6}, \overline{7}, \overline{5}, \overline{a})$ $(1, 3|2, \overline{8}, 7, 6, 5, \overline{9}, \overline{0}, \overline{4}, a)$ $(1, 4|2, \overline{8}, 6, 7, 5, \overline{0}, \overline{9}, 3, a)$

$(1, 5|2, 7, 6, \overline{8}, \overline{a}, \overline{3}, \overline{4}, \overline{0}, \overline{9})$ $(1, 6|2, \overline{5}, \overline{8}, \overline{a}, \overline{3}, \overline{7}, \overline{4}, \overline{0}, \overline{9})$ $(1, 7|2, \overline{5}, \overline{a}, 8, \overline{3}, 6, \overline{4}, \overline{0}, \overline{9})$

$(1, 8|2, 4, 3, 7, a, 6, 5, \overline{9}, \overline{0})$ $(1, 9|2, 6, 7, 5, 4, 0, 3, a, 8)$ $(1, 0|2, 6, 7, 5, 4, \overline{9}, 3, a, 8)$

$(1, a|2, 7, \overline{8}, 6, 5, \overline{9}, \overline{0}, \overline{4}, \overline{3})$ $(2, 3|1, 8, \overline{6}, \overline{5}, \overline{a}, \overline{7}, 0, 4, 9)$ $(2, 4|1, 8, \overline{6}, \overline{7}, \overline{5}, 0, \overline{a}, \overline{3}, 9)$

$(2, 5|1, \overline{6}, 8, a, 3, \overline{7}, 4, 0, 9)$ $(2, 6|1, 7, 5, 8, a, 3, 4, 0, 9)$ $(2, 7|1, \overline{6}, 8, 3, a, 5, 4, 0, 9)$

$(2, 8|1, \overline{4}, \overline{3}, \overline{a}, \overline{6}, \overline{5}, \overline{7}, 0, 9)$ $(2, 9|1, \overline{6}, \overline{7}, \overline{5}, \overline{a}, \overline{3}, \overline{4}, \overline{8}, \overline{0})$ $(2, 0|1, \overline{6}, \overline{7}, \overline{5}, \overline{4}, \overline{a}, \overline{3}, \overline{8}, 9)$

$(2, a|1, 8, \overline{6}, \overline{5}, 3, \overline{7}, 0, 4, 9)$ $(3, 4|1, \overline{9}, \overline{a}, 2, 7, 5, \overline{0}, \overline{8}, 6)$ $(3, 5|1, \overline{9}, \overline{6}, 8, \overline{4}, \overline{0}, a, \overline{2}, \overline{7})$

$(3, 6|1, \overline{9}, 5, \overline{0}, \overline{4}, 8, a, \overline{2}, \overline{7})$ $(3, 7|1, 8, \overline{9}, \overline{4}, \overline{0}, \overline{2}, a, 5, 6)$ $(3, 8|1, 2, \overline{a}, \overline{6}, 4, 0, \overline{5}, 9, \overline{7})$

$$(3,9|1,0,4,\bar{a},2,7,\bar{8},6,5) \quad (3,0|1,\bar{9},\bar{a},2,7,5,4,\bar{8},6) \quad (3,a|1,8,\bar{6},\bar{5},\bar{2},\bar{7},0,4,9)$$

$$(4,5|1,\bar{9},\bar{6},8,3,a,\bar{7},\bar{0},\bar{2}) \quad (4,6|1,7,\bar{9},5,\bar{0},3,a,8,\bar{2}) \quad (4,7|1,\bar{6},\bar{9},8,3,a,5,\bar{0},\bar{2})$$

$$(4,8|1,2,\bar{6},\bar{a},\bar{3},0,\bar{5},\bar{7},9) \quad (4,9|1,\bar{3},\bar{a},2,\bar{8},7,6,5,\bar{0}) \quad (4,0|1,2,5,7,\bar{a},\bar{3},\bar{8},6,9)$$

$$(4,a|1,\bar{6},8,0,\bar{5},\bar{7},\bar{2},\bar{3},9) \quad (5,6|1,\bar{8},\bar{a},\bar{9},3,\bar{4},\bar{0},\bar{2},\bar{7}) \quad (5,7|1,\bar{9},\bar{4},\bar{0},\bar{2},\bar{3},\bar{a},\bar{8},6)$$

$$(5,8|1,7,2,0,4,3,9,a,6) \quad (5,9|1,2,7,\bar{a},\bar{8},6,\bar{3},\bar{4},\bar{0}) \quad (5,0|1,2,\bar{4},7,\bar{a},\bar{3},\bar{8},6,9)$$

$$(5,a|1,7,2,3,0,4,9,\bar{8},6) \quad (6,7|1,\bar{3},\bar{a},\bar{8},\bar{5},2,0,9,4) \quad (6,8|1,7,2,3,a,4,0,9,\bar{5})$$

$$(6,9|1,2,\bar{8},\bar{a},\bar{5},\bar{3},\bar{4},\bar{7},\bar{0}) \quad (6,0|1,2,\bar{8},\bar{a},\bar{3},\bar{4},\bar{5},\bar{7},9) \quad (6,a|1,7,2,3,\bar{8},4,0,9,\bar{5})$$

$$(7,8|1,\bar{a},\bar{6},\bar{5},2,0,4,9,3) \quad (7,9|1,2,\bar{5},\bar{a},\bar{3},\bar{8},\bar{4},6,\bar{0}) \quad (7,0|1,2,\bar{4},\bar{5},\bar{a},\bar{3},\bar{8},6,9)$$

$$(7,a|1,8,\bar{6},\bar{5},3,2,0,4,9) \quad (8,9|1,0,2,4,7,3,5,a,6) \quad (8,0|1,\bar{9},2,7,5,4,3,a,6)$$

$$(8,a|1,5,\bar{9},\bar{0},\bar{4},6,\bar{3},\bar{2},\bar{7}) \quad (9,0|1,\bar{4},\bar{5},\bar{7},\bar{6},\bar{2},8,a,3) \quad (9,a|1,\bar{6},8,\bar{5},\bar{7},\bar{2},\bar{3},\bar{4},\bar{0})$$

$$(0,a|1,\bar{6},8,\bar{4},\bar{5},\bar{7},\bar{2},\bar{3},9)$$

Table 7-1. An oriented matroid admissible with a triangulated manifold of genus 4 with 11 points due to U. Brehm.

This oriented matroid shows that for given genus g the minimal numbers of vertices of combinatorial manifolds and of matroid manifolds also coincide in the case $g = 4$ which was previously unknown. This oriented matroid was also an interesting output (after several CPU hours) of an extended version of the following general purpose algorithm which is formulated here for triangulated manifolds. We call an oriented matroid *admissible* with respect to a triangulated manifold if an affine realization of the oriented matroid would not violate any non-intersection property of an embedding of this manifold, compare also 1.5.

Algorithm 7.3. *Input: Triangulated manifold C with n points, dimension d-1. Chirotope $\chi_0 \in \{-1,0,1\}^{\Lambda(n-1,d)}$ admissible to C. Output: Matroid manifold $\chi \in \{-1,0,1\}^{\Lambda(n,d)}$ of rank d, admissible to C or empty set.*

1. $\chi := \chi_0 \in \{-1,0,1\}^{\Lambda(n-1,d)}$.
2. Construct the set $\mathcal{L}_\chi := \{L_1, L_2, \ldots, L_l\}$ of all oriented hyperlines of χ.
3. Construct the set of contractions $\{\chi/L_1, \chi/L_2, \ldots, \chi/L_l\}$.
4. Construct the sets of possible one point extensions of these contractions χ/L_i:

$$\mathcal{P} := P_{\chi/L_1} \times P_{\chi/L_2} \times \ldots \times P_{\chi/L_l}$$

5. If $\mathcal{P} = \emptyset$, STOP.
6. Pick an extension $ext = (ext_i)_{1 \leq i \leq l} \in \mathcal{P}$.
7. If ext contradicts the oriented matroid property $\mathcal{P} := \mathcal{P} \setminus \{ext\}$, GO TO 5.
8. If ext contradicts the manifold property, $\mathcal{P} := \mathcal{P} \setminus \{ext\}$, GO TO 5.

9. *Output: Matroid manifold* $\chi \in \{-1, 0, 1\}^{\Lambda(n,d)}$ *of rank d admissible to* $\mathcal{C}$,
STOP.

Remark 7.4. *The input triangulated manifold can be replaced by any other input which can easily be checked for oriented matroids. The choice of a possible one point extension of* χ/L_i *reduces the possible one point extensions of* χ/L_j *for* $j > i$, *and moreover, checking the combinatorial properties after each such choice reduces the number of steps substantially. Therefore, we recommend this algorithm for practical purposes only if similar ideas are being implemented.*

There exists a FORTRAN-implementation (J. Bokowski / F. Anheuser) with later changes due to R. Dauster for the uniform case in which these ideas are included. The algorithm is a very sophisticated one, and we don't describe it here. A forthcoming, more general implementation was started in the language C by A. Guedes de Oliveira. It is planned for the non-uniform case as well.

This algorithm was used for finding matroid manifolds in a variety of cases of geometric interest. The group of A. Bachem used it for generating special examples. More applications will be discussed in Chapter VIII.

7.2. On the solvability sequence method
In Chapter II on the existence of algorithms, it was pointed out that the existence of algorithms for deciding the realizability problems of matroids and oriented matroids over real closed fields is equivalent to the existential theory of real closed fields.

In order to find solutions in a reasonable time or in order to find solutions at all, we have to give up the general type of problem. This section is devoted to the realizable case, whereas the next section deals with the dual problem of finding final polynomials.

When an oriented matroid is given and we are looking for coordinates, the dependencies between the signs of bases can be used for reducing the corresponding algebraic system of equations and inequalities. Compare the notion of a reduced system in Section 3.3. Also note that there is some hope of improving these ideas by an algebraic approach described in Chapter IX.

This reduction is of high practical significance, and we assume that such a technique will be applied first. Once such a reduced system is found, the more difficult part begins. This section deals with the question of what can be done not only theoretically but which techniques must be applied for having a chance of solving these systems in real time.

In general, suitable transformations of all variables might be helpful for reducing the system further. Particularly in symmetric cases, we recommend choosing coordinates according to an expected symmetry. Whereas such ingenious techniques might be very useful and whereas creative and intuitive searching is sometimes the only way left, here we are more interested in providing general algorithmic ways, yet leading to solutions in certain general classes of synthetic problems.

A particular subclass of problems was extracted from the following more general idea. It was discussed in the literature as the *solvability sequence method* [31], and we are going to describe some features of this method. But before doing so, we emphasize that some more general ways are possible and when computer algebra methods are used, we recommend an interactive way of applying these ideas.

The general idea reads as follows: Given a set of polynomials, e.g. in our cases determinants with all entries as variables $P_i \in K[x_1, \ldots, x_s]$ with K any ordered field. We make the assumption that we can rewrite these inequalities for at least those polynomials containing a particular variable x_i equivalently in the form

$$L_j(x_1, \ldots, x_{i-1}, x_{i+1}, \ldots, x_s) < x_i \text{ and } x_i < R_k(x_1, \ldots, x_{i-1}, x_{i+1}, \ldots, x_s).$$

We further assume $x_i > 0$ for all i, which is no loss of generality in our cases of determinants. Note also that we have a linear structure at the beginning.

Now we can replace these $j + k$ inequalities by $j \times k$ inequalities, namely $L_i < R_k$ for all i, k thus reducing the number of variables by one.

In some cases combinatorial properties guarantee that this solving for variables is possible in the next step as well, thus making sure that the procedure is successful. This particular case was called the solvability sequence method, [31].

But in general, this straightforward heuristic method can be applied as long as the effort of handling the new system of inequalities does not stop the procedure. The degree of the polynomials occuring later might stop this procedure or the number of inequalities might be too large. The last drawback can be avoided in part by assuming a certain ordering of the inequalities on the left and on the right, and adding these new inequalities describing the ordering and solving the remaining system. In this case the system has a smaller number of inequalities but we are left with a lot of cases. Experience will show which variant is more likely to be successful. Remember that in solving these inequalities, one might even get a contradiction in terms of a final polynomial.

Although these methods appear very elementary, they have proved to be successful even without computer-algebra support.

It was the special case of the solvability sequence method, first implemented for finding coordinatizations for a large class of polytopal spheres, which proved to be successful in deciding several geometric realizability problems. The completion of the enumeration of all neighborly 4-polytopes with 10 vertices [32] was supported by this method, and embeddings of 2-manifolds of higher genus were found in applying variants of this method [20], [21], see also Chapter VIII. It can be shown that all realizations found by the solvability method fulfill the isotopy-property, [31].

We already mentioned that the inequality systems of a reduced system have a very special structure, which sometimes allows describing the algebraic solution process in purely combinatorial terms. A solvability sequence is an ordering of the variables that allows the choice of coordinates in a non-prospective way. It can be shown that the class of realizable oriented matroids with solvability sequences is quite rich although there are examples which do not have this property. Jürgen Richter from Darmstadt gave such an example with 9 points in the plane, [28a].

7.3 On the finding of final polynomials by linear programming

This section is devoted to the more difficult problem of determining final polynomials algorithmically. It is based on a paper of J. Bokowski and J. Richter [27]. We introduce the notion of bi-quadratic final polynomials, and we show that finding them is equivalent to solving an LP-problem. In addition we apply a theorem about symmetric oriented matroids to a series of cases of geometrical interest.

So far, only a rather small number of final polynomials have been found. In every case, these polynomials were determined by using rather long computer-algebra-like calculations, and sometimes even greater effort was needed in solving a specific problem, e.g. in [21], [108]. This more difficult problem of finding a particular final polynomial for specific applications will be discussed. We provide a polynomial algorithmic method, using essentially a suitable variant of linear programming for finding classes of final polynomials. It is very likely that there are final polynomials which we cannot find in this way. On the other hand, the set we do get at least includes all final polynomials known to us so far. Thus our method, deciding first a suitable LP-problem, can be seen as a good pre-processing.

We will deal with symmetrical chirotopes in the Section 7.4. For symmetric d-chirotopes it is natural to decide whether there is a symmetrical realization with respect to a given finite subgroup of the orthogonal group $O(d)$. Our main result in Section 7.4. asserts that, once the non-realizability of our chirotope along with (geometric) symmetry assumptions has been achieved by our method (!), we can solve the general case as well, i.e., the symmetry assumption can be dropped. This is a substantial shortcut in attaining the final polynomial, algorithmically.

It is generally considered that the algorithmic search for final polynomials is very difficult. Sometimes the effort seems to be beyond our reach when the number of points increase. We provide now an algorithmic method for finding final polynomials in a reasonable length of time. In describing the details, we restrict ourselves to the simplicial case, thus simplifying the main idea.

Remember that the change in the Grassmann-Plücker polynomial, caused by a permutation of the points on the right hand side of the vertical line in

$$\{\tau \mid \lambda_1, \lambda_2, \lambda_3, \lambda_4\},$$

does not change the chirotope condition. The polynomial remains equal up to choosing the actual brackets

$$[\lambda_1, \ldots, \lambda_d] \quad \text{or} \quad \text{sign}\pi \, [\lambda_{\pi(1)}, \ldots, \lambda_{\pi(d)}]$$

for a permutation $\pi \in S_d$, where S_d denotes the symmetric group, or it is equal up to ordering the bracket products. By a suitable permutation of the points $\lambda_1, \lambda_2, \lambda_3, \lambda_4$, we can attain that all three bracket products are positive

$$\underbrace{[\tau, \lambda_1, \lambda_2][\tau, \lambda_3, \lambda_4]}_{>0} - \underbrace{[\tau, \lambda_1, \lambda_3][\tau, \lambda_2, \lambda_4]}_{>0} + \underbrace{[\tau, \lambda_1, \lambda_4][\tau, \lambda_2, \lambda_3]}_{>0}.$$

In this case, we say, we have *normalized* our Grassmann-Plücker polynomial. In case χ is realizable and in the normalized case, we can deduce that the inequality

$$[\tau, \lambda_1, \lambda_2][\tau, \lambda_3, \lambda_4] < [\tau, \lambda_1, \lambda_3][\tau, \lambda_2, \lambda_4]$$

must hold. We use a special notation for these *formal inequalities*, arising from normalized Grassmann-Plücker polynomials.

Definition 7.5. *Let χ be a simplicial chirotope and let*

$$\{\tau \,|\, \lambda_1, \ldots, \lambda_4\} := [\tau, \lambda_1, \lambda_2][\tau, \lambda_3, \lambda_4] - [\tau, \lambda_1, \lambda_3][\tau, \lambda_2, \lambda_4] + [\tau, \lambda_1, \lambda_4][\tau, \lambda_2, \lambda_3]$$

be a Grassmann-Plücker polynomial such that

$$\chi((\tau, \lambda_1, \lambda_2)) \cdot \chi((\tau, \lambda_3, \lambda_4)) = \chi((\tau, \lambda_1, \lambda_3)) \cdot$$

$$\chi((\tau, \lambda_2, \lambda_4)) = \chi((\tau, \lambda_1, \lambda_4)) \cdot \chi((\tau, \lambda_2, \lambda_3)) = 1,$$

then we call the 4-tupel BI of formal variables,

$$BI = \langle [\tau, \lambda_1, \lambda_2], [\tau, \lambda_3, \lambda_4] \,|\, [\tau, \lambda_1, \lambda_3], [\tau, \lambda_2, \lambda_4] \rangle,$$

a bi-quadratic inequality of χ. The set of all bi-quadratic inequalities of χ will be denoted by $\mathcal{B}_\chi$.

Whenever χ is realizable, we have

$$A \cdot B < C \cdot D \qquad \text{for all} \qquad \langle A, B \,|\, C, D \rangle \in \mathcal{B}_\chi.$$

In order to prepare a definition for *bi-quadratic final polynomials*, we identify those brackets, which in the realizable case would change according to the alternating determinant rules, i.e., for any permutation $\pi \in S_d$ we set

$$[\lambda_1, \lambda_2, \ldots, \lambda_d] - \text{sign}\,\pi \cdot [\lambda_{\pi(1)}, \lambda_{\pi(2)}, \ldots, \lambda_{\pi(d)}] = 0.$$

In other words, if R is the integer polynomial ring generated by the formal brackets $\{[\lambda] \,|\, \lambda \in \{1, \ldots, n\}^d\}$ and if I is the ideal generated by polynomials of the form

$$[\lambda_1, \lambda_2, \ldots, \lambda_d] - \text{sign}\,\pi \cdot [\lambda_{\pi(1)}, \lambda_{\pi(2)}, \ldots, \lambda_{\pi(d)}],$$

we calculate within the quotient ring R/I. R/I is similar, but not identical, to the bracket ring introduced by White, see [160]. We define *bi-quadratic final polynomials* as follows.

Definition 7.6. *A simplicial chirotope* χ *admits a bi-quadratic final polynomial, whenever, there is a collection of bi-quadratic inequalities*

$$\langle A_i, B_i \mid C_i, D_i \rangle \in \mathcal{B}_\chi; \quad 1 \le i \le k$$

such that the following equality holds within the ring R/I.

$$\prod_{i=1}^{k} A_i \cdot B_i \stackrel{I}{=} \prod_{i=1}^{k} C_i \cdot D_i.$$

Here $a \stackrel{J}{=} b$ denotes that the ring variables a and b are equal modulo the set J.

This definition allows the following claim.

Lemma 7.7. *If* χ *admits a bi-quadratic final polynomial, then* χ *is not realizable.*

Proof of Lemma 7.7.
χ admits a bi-quadratic final polynomial, i.e., there is a collection of bi-quadratic inequalities

$$\langle A_i, B_i \mid C_i, D_i \rangle \in \mathcal{B}_\chi; \quad 1 \le i \le k$$

such that

$$\prod_{i=1}^{k} A_i \cdot B_i \stackrel{I}{=} \prod_{i=1}^{k} C_i \cdot D_i.$$

In assuming that χ is realizable, we have by Remark 1:
$A_i \cdot B_i < C_i \cdot D_i$ for all $i \in \{1, \ldots, k\}$, and since both sides in the above equation are by definition positive, we also have

$$\prod_{i=1}^{k} A_i \cdot B_i < \prod_{i=1}^{k} C_i \cdot D_i.$$

In using

$$[\lambda_1, \lambda_2, \ldots, \lambda_d] - \text{sign}\pi \cdot [\lambda_{\pi(1)}, \lambda_{\pi(2)}, \ldots, \lambda_{\pi(d)}] = 0,$$

whenever neccessary, we arrive at a contradiction.

$\square$

Remark 7.8. *Now, it is a straight-forward task of translating a bi-quadratic final polynomial into an ordinary final polynomial as introduced in Chapter IV.*

Of course, it seems as if one looses quite a bit of information in searching for bi-quadratic final polynomials rather then looking for final polynomials in general. But there are two reasons for doing this.

1. The solvability of the remaining system of inequalities can now be seen to be a LP-problem, yielding a polynomial algorithm by the ellipsoid method, compare e.g. [26].

2. Again all known examples of non-representable oriented matroids, where non-representability was proved by means of final polynomials, can be treated automatically and more easily than before, compare [5], [28a], [24], [27].

And above all, this might be the only accessible method of solving a more difficult problem under consideration. In any case, we suggest solving this LP-problem as a pre-processing method. How do we get the LP-problem?

We consider the set of all formal inequalities $\mathcal{B}_\chi$. We replace each variable $X = [\lambda_1, \ldots, \lambda_d]$ by their formal absolute value $X \cdot \chi(\lambda_1, \ldots, \lambda_d)$. It is clear how one gets new formal inequalities $\langle A, B \mid C, D \rangle$ for the formal absolute values. Assuming realizability, we can take the logarithm on both sides, and we get a linear problem $(*)$ with integer coefficients and strict inequalities for the new variables $Y := log(X \cdot \chi(\lambda_1, \ldots, \lambda_d))$ chosen appropriately.

In principle, it is now possible to use the special structure of this inequality system when solving this system of inequalities. Consider the inequalities with positive coefficients

$$\sum \ldots + \alpha_l Y + \ldots < \sum \ldots + \ldots$$

or

$$\sum \ldots + \ldots < \sum \ldots + \beta_r Y + \ldots \quad ,$$

and multiply those inequalities, which contain the variable Y, with suitable factors in order to get mY, where m denotes the smallest common multiple of all (positive) factors α_l, β_r of Y. We rewrite all inequalities in the form

$$\sum \ldots < mY < \sum \ldots \quad .$$

The decisive inductive step of our process for solving all variables is replacing this system of inequalities by all pairs of inequalities, and comparing all possible parts, left and right that one has solved for Y together with the remaining system of inequalities. Now, either one gets a solution this way, or one finally arrives at a contradiction $0 < 0$, which then can be traced back in order to find the bi-quadratic final polynomial. We formulate this result as follows.

Theorem 7.9.

A chirotope χ admits a bi-quadratic final polynomial, if and only if the dual of the above LP-problem (∗) is admissible (LP-Phase I). Moreover, the solution of the LP-problem can be used for constructing the bi-quadratic final polynomial.

Remark 7.10. *When judging the above theorem, it is the algorithmic point of view one has to keep in mind, the practical significance can be shown in concrete applications.*

7.4. Final polynomials with symmetries

This section is devoted to symmetric chirotopes and corresponding final polynomials with respect to a given symmetry. For symmetric d-chirotopes it is natural, not only to look for a realization at all, but also to decide whether there is a symmetrical realization with respect to a given finite subgroup of the orthogonal group $O(d)$.

This leads us to the tool of *bi-quadratic final polynomials with respect to a symmetry group* for disproving symmetrical embeddings. Our main result in this section asserts that once the non-realizability of our chirotope according to (geometric) symmetry assumptions has been established, by using this tool (!), we can then derive a final polynomial even for the general case, by dropping the (geometric) symmetry assumption. The result being that our chirotope is not realizable at all. The advantage of our theorem is immediate as in these cases, it is an essential shortcut in finding a final polynomial, algorithmically.

In order to formulate this theorem, we first have to fix some notation. Let χ be a simplicial d-chirotope. An element $\sigma \in S_n$ of the permutation group S_n is called a *rotation of* χ if

$$\chi(\lambda) = \chi(\sigma\lambda) \text{ for all } \lambda = (\lambda_1, \ldots, \lambda_d) \in \{1, \ldots, n\}^d,$$

denoted by $\sigma\chi = \chi$. Similarly, an element $\sigma \in S_n$ is called a *reflection of* χ if

$$\chi(\lambda) = -\chi(\sigma\lambda) \text{ for all } \lambda = (\lambda_1, \ldots, \lambda_d) \in \{1, \ldots, n\}^d.$$

In this case, we write $\sigma\chi = -\chi$. We call

$$R_\chi = \{\sigma \in S_n \mid \sigma\chi = \chi\}$$

the *set of rotations of* χ and

$$M_\chi = \{\sigma \in S_n \mid \sigma\chi = -\chi\}$$

the *set of reflections of* χ. The union of these sets forms the group of automorphisms $\text{Aut}(\chi)$ of χ denoted by $\quad G_\chi := R_\chi \cup M_\chi = \text{Aut}(\chi).$

The group G_χ (resp. any subgroup of G_χ) acts on the polynomial ring R as follows. For any bracket and any $\sigma \in S_n$, we define

$$\sigma * [\lambda] = \begin{cases} [\sigma\lambda] & \text{if } \sigma \in R_\chi, \\ -[\sigma\lambda] & \text{if } \sigma \in M_\chi. \end{cases}$$

For any polynomial $P([\lambda], [\mu], \ldots)$ in the bracket variables $[\lambda], [\mu], \ldots$, we define

$$\sigma * P([\lambda], [\mu], \ldots) = P(\sigma * [\lambda], \sigma * [\mu], \ldots).$$

The action of the group can be extended to any quotient R/J, whenever the ideal J remains fixed under the group, in particular, we define for the above polynomial ring R/I : for any $P + I \in R/I$; $P \in R$ and any $\sigma \in G_\chi$:

$$\sigma * (P + I) := (\sigma * P) + I.$$

According to

$$\sigma * \langle A, B \,|\, C, D \rangle = \langle \sigma * A, \sigma * B \,|\, \sigma * C, \sigma * D \rangle,$$

G_χ also acts on the set of bi-quadratic inequalities.

Lemma 7.11. *The property of $\langle A, B \,|\, C, D \rangle$, being a bi-quadratic inequality, remains fixed under the action of G_χ, i.e., for any $\sigma \in G_\chi$, $\sigma * \langle A, B \,|\, C, D \rangle$ is again a bi-quadratic inequality.*

Proof.   If $\langle A, B \,|\, C, D \rangle$ is a bi-quadratic inequality, then there are brackets E and F such that $AB - CD + EF$ is a Grassmann-Plücker polynomial, and $AB > 0$, $CD > 0$, $EF > 0$. Since Grassmann-Plücker polynomials in R/I do not change by renumbering the vertices,

$$(\sigma * A)(\sigma * B) - (\sigma * C)(\sigma * D) + (\sigma * E)(\sigma * F)$$

is a Grassmann-Plücker polynomial as well. For any rotation or reflection σ, we have:

$$(\sigma * A)(\sigma * B) > 0, \quad (\sigma * C)(\sigma * D) > 0, \quad (\sigma * E)(\sigma * F) > 0.$$

Thus, by Definition 2.1, $\langle \sigma * A, \sigma * B \,|\, \sigma * C, \sigma * D \rangle$ yields a bi-quadratic inequality again, in other words, G acts on the set of bi-quadratic inequalities.   □

Now, we introduce the notion of a *symmetric realization of χ with respect to G_R*, where $G_\mathcal{R} < G_\chi$ and $G_\mathcal{R} < O(d)$, is any subgroup of G_χ and $O(d)$ respectively. By that, we mean a realization R of χ with a geometric automorphism group isomorphic to $G_\mathcal{R}$.

Given such a symmetric realization, we have for any $\lambda \in E^d$ and any $\sigma \in G_{\mathcal{R}}$

$$[\lambda] - \sigma * [\lambda] = 0.$$

We will factorize our ring R/I, such that elements together with their symmetric images are identified. Therefore, we define the ideal $I_{G_{\mathcal{R}}}$ in R which is generated by polynomials of the type
$[\lambda] - \sigma * [\lambda]$ for any $\sigma \in G_{\mathcal{R}}$ and any $\lambda \in E^d$. We also consider the polynomial ring $R/\overline{I_{G_{\mathcal{R}}}}$, where $\overline{I_{G_{\mathcal{R}}}}$ denotes the ideal generated by $I \cup I_{G_{\mathcal{R}}}$.

Here we have factored out the symmetry of χ.

Definition 7.12. *A simplicial chirotope χ admits a **symmetric bi-quadratic final polynomial** with respect to $G_{\mathcal{R}}$, whenever, there is a collection of bi-quadratic inequalities*

$$\langle A_i, B_i \,|\, C_i, D_i \rangle \in \mathcal{B}_\chi; \quad 1 \le i \le k,$$

such that

$$\prod_{i=1}^{k} A_i \cdot B_i \overset{\overline{I_{G_{\mathcal{R}}}}}{=} \prod_{i=1}^{k} C_i \cdot D_i.$$

With this definition we obtain the following result:

Theorem 7.13. *For any subgroup $G_{\mathcal{R}} < G_\chi$ we have: A (symmetric) chirotope admits a bi-quadratic final polynomial, if and only if it admits a symmetric bi-quadratic final polynomial with respect to $G_{\mathcal{R}}$.*

Remark: The result of Theorem 7.13. heavily and positively influences the computational part of our applications. Whenever we want to prove that a symmetric chirotope is not realizable and we are looking for a bi-quadratic final polynomial, we can reduce the computational time tremendously. Especially, if the inequality system being considered generally has N inequalities, we only have to solve a system of approximately $N/|G_\chi|$ inequalities. Our proof will show how to construct the bi-quadratic final polynomial whenever the symmetric one is given.

Proof of Theorem 7.13.
Since for any $P, Q \in R$,

$$P \overset{I}{=} Q \text{ also implies } P \overset{\overline{I_{G_{\mathcal{R}}}}}{=} Q,$$

the existence of a bi-quadratic final polynomial implies the existence of a symmetric one. To prove the reverse, we first provide the following two lemmas.

Lemma 7.14. *For any $\lambda_1, \lambda_2 \in E^d$, we have*

$$[\lambda_1] \overset{\overline{I_{G_{\mathcal{R}}}}}{=} [\lambda_2] \quad \Rightarrow \quad [\lambda_1] \in G_{\mathcal{R}} * ([\lambda_2] + I).$$

Proof of Lemma 7.14. $[\lambda_1] =^{\overline{I_{G_{\mathcal{R}}}}} [\lambda_2]$ implies, there exists a $\sigma \in G_{\mathcal{R}}$ such that $[\lambda_1] =^I \sigma * [\lambda_2]$ holds, and this in turn can be written as

$$[\lambda_1] \in \sigma * [\lambda_2] + I = \sigma * ([\lambda_2] + I) \in G_{\mathcal{R}}([\lambda_2] + I).$$

$\square$

Lemma 7.15. *For any $\lambda_1, \lambda_2 \in E^d$, we have*

$$[\lambda_1] \overset{\overline{I_{G_{\mathcal{R}}}}}{=} [\lambda_2] \quad \Rightarrow \quad \prod_{\sigma \in G_{\mathcal{R}}} (\sigma * [\lambda_1]) \overset{I}{=} \prod_{\sigma \in G_{\mathcal{R}}} (\sigma * [\lambda_2]) \ .$$

Proof of Lemma 7.15. For $\lambda_1, \lambda_2 \in E^d$, we assume

$$[\lambda_1] \overset{\overline{I_{G_{\mathcal{R}}}}}{=} [\lambda_2].$$

We consider the orbits $\Omega([\lambda_i]) := \prod_{\sigma \in G_{\mathcal{R}}} (\sigma * [\lambda_i])$, $,i = 1, 2$, of $G_{\mathcal{R}}$ within R generated by $[\lambda_1]$, and $[\lambda_2]$, respectively. They induce corresponding orbits

$$\Omega([\lambda_i] + I) = \prod_{\sigma \in G_{\mathcal{R}}} \sigma * \{[\lambda_i] + I\} = \prod_{\sigma \in G_{\mathcal{R}}} (\sigma * [\lambda_i]) + I, \ i = 1, 2.$$

$\square$

To pursue the proof of Theorem 7.13, let us assume, that χ admits a symmetric bi-quadratic final polynomial with respect to $G_{\mathcal{R}}$. Then, there is a collection of bi-quadratic inequalities

$$\langle A_i, B_i \,|\, C_i, D_i \rangle \in \mathcal{B}_\chi; \quad 1 \le i \le k,$$

such that

(**) $$\prod_{i=1}^{k} A_i \cdot B_i \overset{\overline{I_{G_{\mathcal{R}}}}}{=} \prod_{i=1}^{k} C_i \cdot D_i \,.$$

We will show that in replacing any bi-quadratic inequality by its orbit under $G_{\mathcal{R}}$, we get a bi-quadratic final polynomial in R/I. In other words, the collection

$$\sigma * \langle A_i, B_i \,|\, C_i, D_i \rangle \in \mathcal{B}_\chi; \quad 1 \le i \le k; \quad \sigma \in G_{\mathcal{R}}$$

is a bi-quadratic final polynomial. Finally, we have to show that

$$\prod_{\substack{1 \le i \le k \\ \sigma \in \bar{G}_{\mathcal{R}}}} (\sigma * A_i) \cdot (\sigma * B_i) \stackrel{I}{=} \prod_{\substack{1 \le i \le k \\ \sigma \in \bar{G}_{\mathcal{R}}}} (\sigma * C_i) \cdot (\sigma * D_i).$$

This is equivalent to

$$\prod_{1 \le i \le k} \left(\prod_{\sigma \in \bar{G}_{\mathcal{R}}} \sigma * A_i \right) \cdot \left(\prod_{\sigma \in \bar{G}_{\mathcal{R}}} \sigma * B_i \right) \stackrel{I}{=} \prod_{1 \le i \le k} \left(\prod_{\sigma \in \bar{G}_{\mathcal{R}}} \sigma * C_i \right) \cdot \left(\prod_{\sigma \in \bar{G}_{\mathcal{R}}} \sigma * D_i \right)$$

Since (**) is only valid, if for any bracket $[\lambda]$ on the left of (**), there is a bracket $[\lambda']$ on the right of (**) with $[\lambda] = {}^{I_{G_{\mathcal{R}}}} [\lambda']$, the result follows according to Lemma 7.15.

$\square$

Applications of all theorems of the last two sections can be found in the paper [27] which contains also the proofs which were given here. But we pick a particular example for showing the extremely short non-realizability proof-method.

Example 7.16. An oriented Vamos matroid.
We give a non-realizability proof for an oriented Vamos matroid of 8 points in rank 4, compare [124].

As Ziegler pointed out, it is different to the one considered in [109]. To give a combinatorial description of the oriented matroid, we start with the chirotope χ which corresponds to the following rank 4 configuration.

$$\begin{array}{c} 1 \\ 2 \\ 3 \\ 4 \\ 5 \\ 6 \\ 7 \\ 8 \end{array} \left(\begin{array}{cccc} 1 & 1 & 1 & 1 \\ 1 & 1 & -1 & -1 \\ 1 & -1 & 1 & -1 \\ 1 & -1 & -1 & 1 \\ -1 & 1 & 1 & 1 \\ -1 & 1 & -1 & -1 \\ -1 & -1 & 1 & -1 \\ -1 & -1 & -1 & 1 \end{array} \right)$$

Notice that this configuration has a symmetry group isomorphic to $Z_2 \times S_4$, since the points $1, 2, 3$ and 4 form a regular tetraedron in the subspace

$$\{(w, x, y, z) \mid w = 1\}$$

and the points $5, 6, 7$ and 8 form a similar one in the subspace

$$\{(w, x, y, z) \mid w = -1\}.$$

Notice furthermore that we have

$$[1,2,5,6] = [1,3,5,7] = [1,4,5,8] = [2,3,6,7] = [2,4,6,8] = [3,4,7,8] = 0.$$

We modify the chirotope χ in order to get a new map χ_V by replacing exactly these zero-orientations by 1, i.e., we require

$$\chi_V(1,2,5,6) = \chi_V(1,3,5,7) = \chi_V(1,4,5,8) =$$

$$\chi_V(2,3,6,7) = \chi_V(2,4,6,8) = \chi_V(3,4,7,8) = +1$$

and $\chi = \chi_V$ for all other 4-tupels. We will show that χ_V is again a chirotope. We consider any Grassmann-Plücker polynomial $\{a,b \mid \dots\}$ where a,b are vertices of the same tetrahedron, as mentioned above. In this polynomial $\{a,b \mid \dots\}$, there occurs no other basis λ with $\chi(\lambda) = 0$, since all 4-tupels contain a and b. But if a and b lie in different tetrahedra, and if there would be another λ in the polynomial with $\chi(\lambda) = 0$, we see from the structure of the polynomial $\{a,b \mid \dots\}$ and from the six 4-tupels above that this zero-valued $\chi(\lambda)$ is the second factor in the same summand of the polynomial which is already equal to zero. Two other summands of opposite signs must exist, showing that the chirotope condition for χ_V is fulfilled.

Notice that χ_V has a combinatorial symmetry isomorphic to the alternating group A_4 which can be generated by the permutations

$$(123)(567) \text{ and } (12)(34)(56)(78).$$

So far, we have only given a description of our chirotope χ_V. Now we claim that χ_V is not realizable.

Proposition 7.17. *The Vamos matroid χ_V is not realizable, and χ_V admits a bi-quadratic final polynomial.*

Proof Consider the Grassmann-Plücker polynomial

$$\{1,2 \mid 3,6,5,4\} =$$
$$= \underbrace{[1,2,3,6]}_{+}\underbrace{[1,2,5,4]}_{+} - \underbrace{[1,2,3,5]}_{+}\underbrace{[1,2,6,4]}_{+} + \underbrace{[1,2,3,4]}_{-}\underbrace{[1,2,6,5]}_{-} = 0.$$

$\langle [1,2,3,6], [1,2,5,4] \mid [1,2,3,5], [1,2,6,4] \rangle$ is a bi-quadratic inequality of χ_V. We assume that χ_V is symmetrically (A_4) realizable, therefore, we have permutations $\sigma_1 := (123)(567),\quad \sigma_2 := (124)(568) \in G_{\chi_V}.$ such that

$$\sigma_1^{-1} * [1,2,3,6] = [1,2,3,5] \text{ and } \sigma_2 * [1,2,5,4] = [1,2,6,4].$$

Thus, we have $[1,2,3,6][1,2,5,4] =^{I_{G_{\chi_V}}} [1,2,3,5][1,2,6,4]$. In conclusion, Theorem 2 proves the lemma, i.e., χ_V is not realizable at all.

$\square$

Example 7.18. Classification of matroid manifolds.

It was a classification problem for 31 matroid manifolds what actually stimulated these investigations and which led to these results. These results will be given and described in [28].

COMPUTATIONAL SYNTHETIC GEOMETRY

INPUT: combinatorial or geometric condition
Felix Klein's regular map $\{3, 7\}_8$

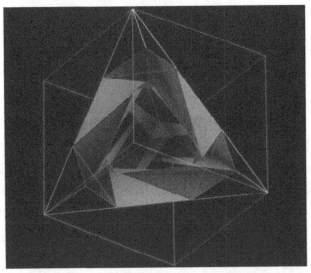

OUTPUT: coordinates or non-realizability proof
Polyhedral realization of Felix Klein's map
E. Schulte and J.M. Wills (1986).

Chapter VIII

ON THE FINDING OF POLYHEDRAL MANIFOLDS

There is a long tradition in studying polyhedral manifolds and complexes including the special case of convex polyhedra [76]. We look at this field from the computational synthetic geometry point of view. Or to put it the other way round, in this chapter we look at computational synthetic geometry from the point of view of studying polyhedral manifolds and some generalizations. Several aspects came about and methods were developed while problems were being studied in convexity. In the meantime, methods which have been proved successful in studying such geometric problems have been carried over and enriched to the more general setting of computational synthetic geometry.

This short history is written by the first author and reviewed in Section 8.1. In Section 8.2 we mention some combinatorial construction techniques which provide us with combinatorial input for problems in computational synthetic geometry, and in Section 8.3 we report about results in which the systematic approach suggested in this monograph has made challenging contributions. Particular interest is devoted to questions of symmetry, such as problems connected to regular polyhedra with hidden symmetries.

Whereas intuition sometimes helps in finding realizations as in the case of Felix Klein's map or Dyck's map, [136], [39], [35], in this monograph the more systematical or algorithmical aspect is suggested. Nevertheless, the result depicted on the opposite page stands as an impressive example of remembering the general definition of *Computational Synthetic Geometry* given in Chapter I. For more details about regular maps and in particular about the automorphism group of Felix Klein's map, we refer to [35], [73].

We thank Jörg Witzel and the Computer Graphics Center, Darmstadt for support in getting these pictures.

8.1. Historical remarks

It seems difficult to think back to the time when oriented matroids were not known to be a fundamental tool in a variety of different fields of mathematics, and although first results were written up in the literature, the structure was not accepted in the way it is seen today. It still happens that other authors are found having studied this structure without being aware of developments in other areas of research. The paper of Buchi and Fenton [40a] can be viewed as such an example. See also our remarks about oriented matroids in Section 1.

The structure of oriented matroids was used by the first author much later than other authors but has been independently used since 1979. At that time Grünbaum & Sreedharan [82], Altshuler [3], [5], and other authors had given complete classifications for 4-polytopes with few vertices by first enumerating all spheres and then trying to decide their polytopality. There were several geometric methods which were applied and tested for these problems and among others there was a strong wish in Ewald's school to decide them. A particular case in Altshuler's list was left open, and at that time several authors did not succeed in solving this problem with their methods. At that time, Bokowski gave a short proof which solved this last open case, and moreover, the actual proof was extremely short. It was the combinatorial and algebraic reduction technique that had led to solving this case by hand and which was not published at that time. The boring calculations seemed not to be of much interest.

Some time later Kleinschmidt found an interesting sphere [23], and he tried to decide whether it is polytopal, compare Chapter VI, Theorem 6.8, and Chapter IV, after Example 4.23. When Kleinschmidt failed to find coordinates, Bokowski decided to apply his method again. But this time he looked for as much computational support as possible. Programs were written at that time by J. Bokowski, C. Antonin, and B. Neidt in order to decide Kleinschmidt's sphere with this new method. This approach was finally successful when Ewald claimed to have a proof for this particular case as well. It turned out later that Ewald's geometric proof was correct as well, and the result was written up in [23]. Later on, still another proof was given by Altshuler [4] for this interesting sphere.

But again, Bokowski's method of deciding this example was not written up and not much appeared in the paper [23]. Only sketches were given. The method had the advantage that all calculations were not needed in the actual proof, e.g. the proof could be given without describing the methods in detail. The motivation of investing much time in deciding more small examples, and to get computational support was difficult. It was planned to test this method in other cases in more detail. It was the interesting case M_{963}^{10}, given by Altshuler and investigated in the paper [24], compare also Section 5, which led to a new non-realizability proof like in the case M_{963}^9. Not much later another interesting case, (Perles posed it again as a problem), sphere M_{425}^{10} in Altshulers list, was decided by additional hand calculations. There was the well-founded hope that now all of the 99 cases open in Altshuler's list of spheres at that time could be decided.

The method was then applied and tested for deciding a larger class of spheres which was successful and which was published in [32]. At that time we learned from Dress that it was the structure of oriented matroids that was always used in the reduction of the Steinitz problem. This stimulated our research very much, and now it was accepted to write up and publish what could be applied in a much broader setting. Even small examples which could be handled without much effort compared with the 4-dimensional former cases were studied. Most of the development afterwards, its applications and its connections to other areas of research, are the content of this monograph.

It is this perspective which led to choosing the algorithmic Steinitz problem *"Given a lattice L, is it polytopal?"* as the first problem in computational synthetic geometry.

A documentation of some early results was written up by Antonin, [7].

In the meantime, the major part of a system of programs, first thought of supporting the sphere decisions only, was written by B. Sturmfels in joint work with J. Bokowski, [34], taking into consideration all experience of the foregoing parts which had been developed so far. Some new ideas came up during the preparation of the papers [31] and [28a] in which the main part of the method was published for the first time, see also [33].

The inductive search for admissible simplicial oriented matroids for given combinatorial complexes was implemented by F. Anheuser and the first author. A special graphic software program, especially useful in cases of representing the shape of polyhedra, and remarkable for the underlying idea of its hidden surface algorithm, was written by J. Richter. It was developed further by F. Anheuser, using the equipment of the "Computer Graphic Center" in Darmstadt.

Additional programs were written by Anselm Eggert during the preparation of our joint paper [22]. A variant for handling symmetric cases was finished by Ronald Dauster. It is this system of programs which supported much of the results presented in this monograph. For a first report about these programs and its applications, see [34].

8.2. Combinatorial input for generating polyhedral manifolds

The input of a *synthetic* computational geometry problem consists of a combinatorial or geometric condition. Very often these conditions were studied before in a purely combinatorial context, and very often interesting combinatorial objects led to the question of whether a corresponding geometric one exists. We mention these theories as they provide us with challenging input for problems in computational synthetic geometry.

For instance, it was the work of Daublebsky and others [51], [52] which gave complete classifications for certain n_3 configurations, compare Chapter III. It was the work of Grünbaum & Sreedharan [82], Altshuler [3], [5], and other authors which led to complete enumerations of a certain list of 3-spheres as a base for finding out which of them is polytopal. The algorithmic Steinitz problem was already discussed in Chapter I.

In trying to find new regular polyhedra, we have another class of examples which is closely related: It is the approach of looking at analogues of the Platonic solids and of determining those combinatorial 2-manifolds without boundary, of a given genus g, which admits a flag-transitive group of automorphisms. In searching algorithmically for suitable polyhedral realizations, i.e. (in case of $\mathbf{R}^3$) in looking for a finite set of plane polygons whose union (without self-intersections) corresponds to the combinatorial manifold, we solve a problem in computational synthetic geometry. There is a classical theory for determining our combinatorial input, e.g. combinatorial manifolds (regular maps). They are known up to genus 6, [67], and some of them are such famous examples as Dyck's map $\{3,8\}_6$ and Klein's map $\{3,7\}_8$ of genus 3. More details and references can be found in the article [35] and in papers cited there.

New combinatorial manifolds were found, studied, and investigated by Altshuler, Brehm, Kühnel, Lassmann, and others, [21]. The following combinatorial input is due to U. Brehm, see also [20].

Example 8.1. Take a regular tetrahedron. Label the vertices and label the midpoints of the edges as in Figure 1.

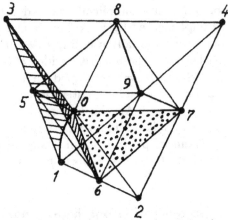

Figure 8-1. A polyhedron of genus 3 with minimal number of vertices.

The convex hull of these new points form an octahedron. Now reduce this octahedron a little bit. Consider the following triangles and all triangles you get under the tetrahedral symmetry A_4. **INPUT DATA:** *(1,3,0), (3,0,6), (0,6,7);* A_4

This provides us with the combinatorial input. We have the following list of triangles.

$1,3,5$	$2,3,8$	$2,5,8$	$1,4,8$
$3,4,0$	$3,8,0$	$1,2,0$	$2,4,5$
$4,5,0$	$1,3,7$	$3,6,7$	$2,3,6$
$2,6,0$	$5,6,0$	$1,5,6$	$1,4,6$
$4,6,9$	$3,4,9$	$3,5,9$	$5,8,9$
$1,8,9$	$1,2,9$	$2,7,9$	$2,4,7$
$4,7,8$	$7,8,0$	$1,7,0$	$6,7,9$

Does a polyhedron without self-intersections with an isomorphic face-structure exist? What is the highest geometric symmetry that the polyhedron can have? These questions formed a challenging example for testing our methods as described in Chapter VII. A solution was published in [20]. When you count the number of different faces, you get by Euler's formula for the genus g of this combinatorial complex $g = 3$.

U. Brehm found another combinatorial manifold depicted in Figure 8-2. This was an interesting case in view of Problem 1.5. In determining all oriented matroids admissible to this manifold and in finding final polynomials for all of them, we would have confirmed this conjecture. But the opposite was true [21], we will discuss this example in Section 8.3.

More than 15 years ago, Duke and Grünbaum worked on Conjecture 1.6: *Every triangulated torus can be embedded in* $\mathbf{R}^3$. In the meantime several other geometers tried to solve this problem. What has been done so far is to make a complete list of all unshrinkable tori and also to provide a corresponding realization in each case. This was done by B. Grünbaum and R.A. Duke and much later independently by S. Lavrenchenko. It could not be established in its full generality that the reversal of the shrinking process could also be performed geometrically. P. Mani-Levitska (private communication) has solved some cases some time ago (unpublished). In the next section, we discuss a particular case, and we study it in order to get a complete overview about all possible (symmetric) realizations. This might open a new way of looking at this problem.

A particular challenging problem is to find a 2-manifold such that the 1-skeleton, the set of all edges, forms a complete graph. A tetrahedron and a realization of Möbius torus [22], the first one given by Császár [50], are the only known realized 2-manifolds with this property. In the next possible case, corresponding to a complete graph with 12 vertices, K_{12}, we have to look for a manifold of genus 6. In Ringel's book two such combinatorial manifolds are given, and Altshuler constructed many more. But so far all attempts have failed to find a corresponding admissible oriented matroid. Thus this problem must remain open. But there is a well-founded hope that future investigations can solve this challenging example.

8.3. Manifolds with minimal number of vertices

2-manifolds with a minimal number of points are viewed as candidates for disproving Conjecture 1.5, compare Section 8.2. Thus they are not only first examples for testing our algorithmic methods in computational synthetic geometry but they might solve a long standing conjecture in geometry, and moreover, they might be of interest in their own right.

Applications of our methods in this area of research have proved to be useful. Consider for instance the input from Example 8.1. We show the method of constructing coordinates: in addition we require geometric symmetries. These methods can be applied in other cases as well, and indeed, in the examples chosen in Chapter III, this has been done when the oriented matroids have already been known and given. In the case of Möbius' torus a complete overview of all symmetric realizations was possible [22].

Let us summarize and recall from previous chapters the main steps in solving such realizability problems. We chose the particular case as depicted in Figure 8.1.

Algorithm
INPUT: *(1,3,0), (3,0,6), (0,6,7); A_4*
OUTPUT: *$P_1, \ldots, P_{10} \in R^3$ (symmetry) or "there is no realization with the given symmetry". (final polynomial)*
1. Find all admissible oriented matroids corresponding to the given input complex.
2. If there is no such admissible oriented matroid, then print *C is not realizable* and STOP.
3. Pick a corresponding oriented matroid.
4. Reduce the system of inequalities.
5. If the corresponding oriented matroid has a solvability sequence (compare [31]), then choose coordinates, print *C is realizable* and STOP.
6. In case a final polynomial p can be found, print *C is not realizable ; final polynomial: p* and STOP.
7. If there is no final polynomial p (this is difficult to decide in practice, although it is possible by Collin's method [43]), print *C is realizable* and STOP.

Example 8.1 has been decided as being realizable [28]. There are even symmetric realizations.

Before investigating a realizable manifold, we consider an example in which non-realizability has already been seen after the second step, and the proof can briefly be given. The example has independently been investigated by A. Altshuler and P. Engel. Their results did not coincide. Altshuler has asked that the result be checked with the techniques we discuss here. We have been given a 3-sphere with 9 points, and we want to decide if it is polytopal. The facets are as follows

```
1235 1236 1245 1248 1268 1345 1347 1367 1478
1679 1689 1789 2357 2367 2456 2468 2569 2579
2679 3458 3478 3579 3589 3789 4569 4589 4689
```

It is sphere No. 24 according to Altshuler's list [2]. We pick a subset of facets. An x in the following matrix indicates incidences between facets and simplices. The orientations of the simplices listed in the left column have to be as marked modulo reversing all of them.

	1235	1245	1347	1345	1478	2357	2367	2456	2468	3478
−24567								x		
−12456		x						x		
−12457		x								
−12345	x	x								
+23457						x				
−23467							x			
−23567						x	x			
+12357	x					x				
−24568								x	x	
+24678									x	
−12345	x			x						
+13457			x	x						
−13478			x		x					x
−34678										x
+14678					x					
+13467			x							

Now the following Grassmann-Plücker-relations have been used: There is no corresponding oriented matroid, e.g. the sphere is not polytopal. We assume realizabilty and find:

The polynomial $\{467 \mid 1238\}$ determines sign[12467] to be -1.

$$\{467 \mid 1238\} = \underbrace{[12467]}_{*}\underbrace{[34678]}_{-} - \underbrace{[13467]}_{+}\underbrace{[24678]}_{+} + \underbrace{[14678]}_{+}\underbrace{[23467]}_{-} = 0$$

The polynomial $\{257 \mid 1346\}$ determines sign[12567] to be 1.

$$\{257 \mid 1346\} = \underbrace{[12357]}_{+}\underbrace{[24567]}_{-} - \underbrace{[12457]}_{-}\underbrace{[23567]}_{-} + \underbrace{[12567]}_{*}\underbrace{[23457]}_{+} = 0$$

The polynomial $\{126 \mid 3457\}$ yields the contradiction (1236 is a facet):

$$\{126 \mid 3457\} = \underbrace{[12346]}_{+s}\underbrace{[12567]}_{+} - \underbrace{[12356]}_{+s}\underbrace{[12467]}_{-} + \underbrace{[12367]}_{-s}\underbrace{[12456]}_{-} = 0$$

□

The following example has been investigated by R. Dauster. It is one of the 21 unshrinkable tori which were mentioned in the last section. The list of all triangles can be seen in Figure 8-3. They are also listed below:

```
1 2 9    1 4 9    2 3 9   3 8 9   1 3 4   3 4 8
4 5 6    4 6 9    6 7 8   6 8 9   4 5 7   4 7 8
1 2 5    2 5 6    2 3 7   2 6 7   1 3 5   3 5 7
```

Why do we study this example? There are several solutions for such embeddings. But in this case we get a complete overview about all symmetric realizations, and it might be useful to start with a particular geometric realization when applying an inductive geometric argument for Problem 1.5.

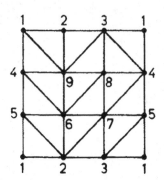

Figure 8-3 Combinatorial manifold of an unshrinkable torus.

The only combinatorial automorphism for this list of triangles is given by the permutation $(1,8)(2,6)(3,4)(5,7)(9)$.

We first have to find an admissible oriented matroid. The corresponding method has been described in Chapter VII. In addition, we are looking for a symmetric(!) realization of this torus. R. Dauster (Diplom-thesis, Darmstadt) found a large list of admissible symmetric affine uniform oriented matroids.

We pick an example from this list and show the way in which coordinates have been found provided they exist. The oriented matroid is given below. All bases are ordered lexicographically, e.g. sign[1,2,3,4]=+, sign[1,2,3,5]=+, ... etc.

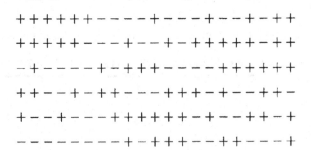

Figure 8.2 Oriented matroid admissible to the torus in Figure 8-3.

The signs of these bases determine corresponding inequalities for the 4×4-sub-determinants of the matrix of homogeneous coordinates. In the realizable case and after a projective normalization, we can choose these homogeneous coordinates for the 9 points as follows:

$$
\begin{pmatrix}
1 & 0 & 0 & 0 \\
0 & 1 & 0 & 0 \\
a & b & c & d \\
e & f & g & h \\
i & j & k & l \\
0 & 0 & 1 & 0 \\
m & n & o & p \\
0 & 0 & 0 & 1 \\
-1 & 1 & 1 & -1
\end{pmatrix}
$$

The next step is then to determine a reduced system. As a reduced system, again with the required symmetry(!) and together with all 1×1 determinants, we have the following:

$a < 0 \quad b > 0 \quad c < 0 \quad d > 0$
$e > 0 \quad f < 0 \quad g > 0 \quad h < 0$
$i > 0 \quad j > 0 \quad k > 0 \quad l < 0$
$m < 0 \quad n > 0 \quad j > 0 \quad p > 0$
$ct - ds < 0 \quad gl - hk > 0 \quad gt - hs < 0 \quad df - bh < 0$
$bg - cf < 0 \quad bp - dn > 0 \quad kr - js < 0 \quad or - ns > 0$
$ah - de < 0 \quad ce - ag < 0 \quad al - di < 0 \quad gi - ek > 0$
$hm - ep < 0 \quad lm - ip < 0$
$-b(k + l) + c(j + l) + d(j - k) < 0$
$-a(gl - hk) + c(el - hi) - d(ek - gi) > 0$
$e(k + l) - g(i - l) - h(i + k) < 0$
$a(fp - hn) - b(ep - hm) - d(fm - en) < 0$
$a(p + n) + b(p - m) - d(m + n) > 0$
$e(o - n) + f(o + m) - g(n + m) < 0.$

For simplicity, we replace all negative variables x by $-x$. The symmetry requirement reduces the number of variables down to 8. The system now looks as follows:

$$a < b \quad a < d \quad b < c \quad c < d$$

$$k < j \quad l < i \quad ab < cd \quad al < di$$

$$dk < bi \quad bl < ak$$

$$aci + abl + ddk < cdl + bdi + aak$$

$$ai + ak + dk < bi + bl + dl$$

$$dj + cl + bl < dk + cj + bk$$

A first observation shows that all inequalities containing the variable i can be omitted. i can always be choosen large enough in order to fulfil these inequalities provided the remaining system of inequalities holds.

1. $i \rightarrow$ infinity

In the remaining system, l is only bounded from below. Therefore, we now assume l to be zero, and later on, we choose l according to the bounds given by the inequalities in which the variable l occur.

2. $l \rightarrow$ zero.

The same argument can now be applied in case of the variable a. The system of inequalities can again be reduced.

3. $a \rightarrow$ zero.

Now all remaining variables are bounded from above or from below. We can solve for c, e.g. we add all pairs of left bounds and right bounds for the variable c to our system of inequalities, and we delete those inequalities containing the variable c.

4. solve for c.

In the remaining system, we see that we can choose the last variables as follows:

5. choose b, d, k, j according to $0 < b < d$ and $0 < k < j$

This is the starting point when we now determine coordinates. We work as follows in reverse:

5. choose b, d, k, j according to $0 < b < d$ and $0 < k < j$

4. choose c according to:

$$b < c < d$$

$$(dj - bk - dk)/j < c$$

3. choose a according to:

$$0 < a < b$$

2. choose l according to:

$$0 < l < (dk + cj + bk - dj)/(b + c)$$

$$0 < l < ak/b$$

1. choose i according to:

$$(abl + ddk - cdl - aak)/(bd - ac) < i$$

$$(ak + dk - bl - dl)/(b - a) < i$$

$$l < i$$

$$dk/b < i$$

We have found a solvability sequence in the symmetrical case. The corresponding realization space for all symmetrical realizations is contractible, compare [31].

□

R. Dauster has written an additional program to connect the computer algebra system MACSYMA with the system of programs which determines admissible oriented matroids for given combinatorial manifolds. An interactive handling of the above inequality systems in case of more difficult problems has been shown to be of much help in practical decisions.

Intuition sometimes helps in finding realizations as in the case of Felix Klein's map or Dyck's map, [136], [39], [18], [35], in this monograph the more systematical or algorithmical aspect has been suggested. The last success in this direction was in finding a polyhedron of genus 4 with minimal number of vertices. We have already listed the oriented matroid in Chapter VII. In Figures 8-6, 8-7 we illustrate the shape of the realization which corresponds to the oriented matroid in Table 7-1. The height of each point is seen on the scale between the last two pictures, and the shape can be studied by constructing the complete polyhedron step by step. To confirm the combinatorial structure it is depicted in Figure 8-5.

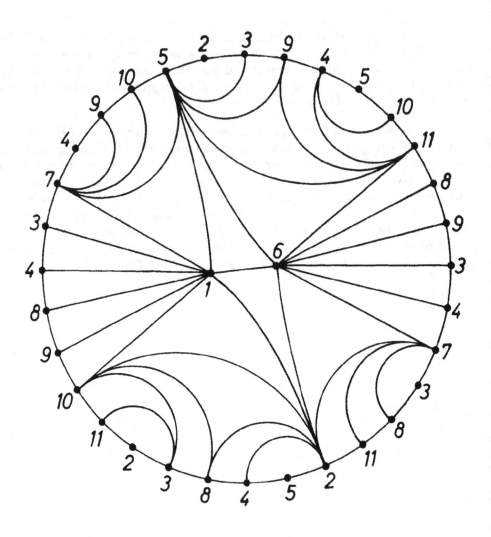

Figure 8-5. Combinatorial manifold of genus 4 with 11 points (Table 7-1).

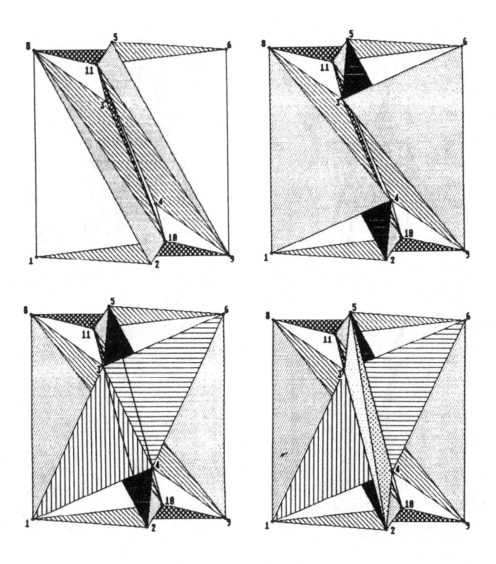

Figure 8-6. Polyhedron of genus 4 with minimal number of vertices, Part 1.

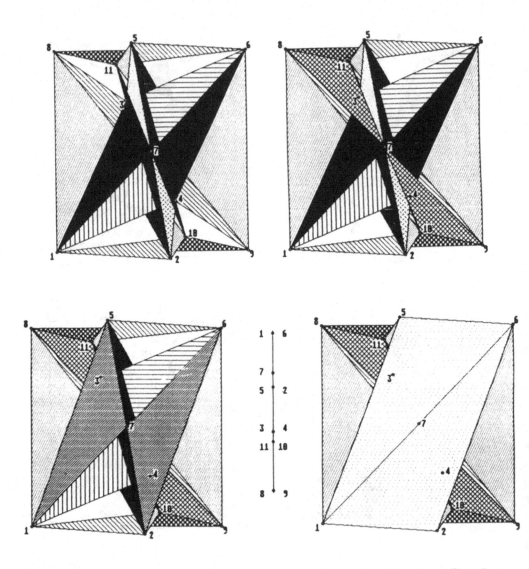

Figure 8-7. Polyhedron of genus 4 with minimal number of vertices, Part 2.

Chapter IX

MATROIDS AND CHIROTOPES AS ALGEBRAIC VARIETIES

We have seen that matroids and oriented matroids play a key role in computational synthetic geometry. We are mainly concerned with the corresponding realizability problems. But very often the given synthetical (geometric computational) problem has to be reduced, and in order to simplify our problem, we have to work within the combinatorial structure of matroids and oriented matroids. Thus all tools for handling these structures are of interest in computational synthetic geometry.

There are several ways of defining matroids and oriented matroids. The context they are used in is very often decisive, and fruitful results can be attained by transferring the results from one side to the other. In this chapter we suggest considering oriented matroids and matroids from an algebraic point of view. Algebraic tools have been further developed, and their application will provide us with new methods within our theory. We are convinced that this algebraic approach is still very promising. Our description is based on the forthcoming paper of J. Bokowski, A. Guedes de Oliveira, and J. Richter [26]. It provides us with new insights, and algebraic investigations will give an impetus to the theory of oriented matroids.

In this chapter we propose an encoding of *oriented matroids as algebraic varieties over a field of characteristic three*. As an application we obtain a straightforward proof for the fact that the three-summand Grassmann-Plücker conditions essentially suffice to define oriented matroids in terms of chirotopes. This approach is of considerable importance for computational purposes because it simplifies a number of fundamental combinatorial algorithms. So far the situation was unsatisfactory because the classical proof for this reduced characterization of chirotopes was based on the following steps.

First one has to apply a proof for the equivalence of both definitions, the one for oriented matroids given by Las Vergnas and Bland [17], and the definition of a chirotope, given by Novoa [121a] and Dress [59]. This proof of equivalence is considered to be a deep result, see [64], [59]. (In fact this is one reason why oriented matroids were being studied at different places without realizing that these structures were being studied in other fields at the same time.) An additional part of the proof is due to Las Vergnas, written in terms of the classical notion of oriented matroids. In glueing these parts together, the desired assertion is yielded when everything is carried over to the chirotope notation. Again, we will show that all this can now be avoided, and moreover, the algebraic approach seems to be straightforward and also useful for similar results.

Combinatorial dependencies can be studied in applying computer algebra tools like Buchberger's Gröbner bases method.

9.1. New algebraic methods for oriented matroids

Before we introduce the algebraic concept for oriented matroids, a similar setting will be described for matroids. In Chapter 1, Definition 1.10, we gave the definition of a *matroid* M as a pair $(E, \mathcal{B})$ where E is a finite set and $\mathcal{B}$ is a collection of subsets of E, called *bases* of M such that for all $B, B' \in \mathcal{B}$ and $e \in B \setminus B'$, there exists $e' \in B' \setminus B$ such that $B \setminus \{e\} \cup \{e'\} \in \mathcal{B}$. This property is an abstract form of Steinitz' exchange lemma.

We consider the set of ordered d-tuples $\Lambda(n,d) = \{[1, 2, \ldots, d], \ldots\}$ also as the set of variables in the polynomial ring $GF_2[\Lambda(n,d)]$ over the field GF_2 of characteristic 2. Any function $f : \Lambda(n,d) \to GF_2$, extends uniquely to an alternating map $f : E^d \to GF_2$ via $f[\lambda_1, \ldots, \lambda_d] = f[\pi(\lambda_1), \ldots, \pi(\lambda_d)]$ and to a ring homomorphism $GF_2[\Lambda(n,d)] \to GF_2$.

Given a matroid $M = (E, \mathcal{B})$, we can define a function $f : \Lambda(n,d) \to GF_2$ such that $\mathcal{B} = f^{-1}(1)$, in other words, $f[\lambda_1, \ldots, \lambda_d] = 1$ if and only if $\{\lambda_1, \ldots, \lambda_d\} \in \mathcal{B}$. ($f$ corresponds to M in Remark 1.11.)

The algebraic formulation comes in now as follows. For a given matroid $M = (E, \mathcal{B})$, we have for any pair of bases $(B = \{b_1, \ldots, b_d\}, B' = \{b'_1, \ldots, b'_d\},)$ and any i, $1 \leq i \leq d$,

$$f(B) f(B') \prod_{k=1}^{d} (1 - f(B \setminus \{b_i\} \cup \{b'_k\})) = 0.$$

On the other hand, given an arbitrary map $f : \Lambda(n,d) \to GF_2$ and requiring this last equation to hold for all pairs

$$\{[b_1, \ldots, b_d], [b'_1, \ldots, b'_d]\} \in \Lambda(n,d)^2$$

and all i, $1 \leq i \leq d$, then the corresponding $M := (E, \mathcal{B})$ will be a matroid.

In other words, the above equations define an algebraic variety over the field GF_2 which can be considered as representing all matroids. In order to formulate this in ideal-theoretical terms, we define for $B = [b_1, \ldots, b_d]$ and $B' = [b'_1, \ldots, b'_d]$ the polynomials $P_{B,i,B'} \in GF_2[\Lambda(n,d)]$,

$$P_{B,i,B'} := [b_1, \ldots, b_d][b'_1, \ldots, b'_d] \prod_{k=1}^{d} (1 - [b_1, \ldots, b_{i-1}, b_{i+1}, \ldots, b_d, b'_k])$$

and the matroid ideal $\mathcal{J}(n,d)$ generated by all these polynomials,

$$\mathcal{J}(n,d) := < \{P_{B,i,B'} \mid B, B' \in \Lambda(n,d), 1 \leq i \leq d\} >$$

Theorem 9.1. (Steinitz' exchange lemma) *A map $f : \Lambda(n,d) \to GF_2$ is a matroid if and only if $f \in \mathcal{V}(\mathcal{J}(n,d))$.*

Definition 9.2. *We call* $\mathcal{V}(n,d)) := \mathcal{V}(\mathcal{J}(n,d))$ *the matroid variety in rank d for n points.*

In a similar fashion we are led to define an algebraic variety $\mathcal{V}^O(n,d)$ over the Galois field GF_3 which represents precisely the set of chirotopes for a given rank and a given number of points. This will be prepared as follows.

For any $\lambda \in \Lambda(n, d+1)$ and any $\mu \in \Lambda(n, d-1)$, we consider the Grassmann-Plücker polynomials (compare Theorem 1.8.)

$$\{|\mu|\lambda\} := \sum_{i=1}^{d+1}(-1)^i[\lambda_1, \ldots, \lambda_{i-1}, \lambda_{i+1}, \ldots, \lambda_{d+1}][\mu_1, \ldots, \mu_{d-1}, \lambda_i]$$

as elements of $GF_3[\Lambda(n,d)]$. The monomials in these Grassmann-Plücker polynomials,

$$M_i := (-1)^i[\lambda_1, \ldots, \lambda_{i-1}, \lambda_{i+1}, \ldots, \lambda_{d+1}][\mu_1, \ldots, \mu_{d-1}, \lambda_i] \in GF_3[\Lambda(n,d)],$$

are used for defining their k-th elementary symmetric function

$$P^{(k)}_{\{|\mu|\lambda\}} := e^{(k)}(M_1, \ldots, M_{d+1}) := \sum_{1 \leq i_1 < \ldots < i_k \leq d+1} M_{i_1} \ldots M_{i_k}.$$

The sum over all such polynomials with odd, or even degree k will be denoted by

$$od\{|\mu|\lambda\} := \sum_{k \, odd} P^{(k)}_{\{|\mu|\lambda\}}, \quad or \quad ev\{|\mu|\lambda\} := \sum_{k \, even} P^{(k)}_{\{|\mu|\lambda\}},$$

respectively. $\mathcal{J}^O(n,d)$ denotes the ideal generated by all polynomials $od\{|\mu|\lambda\}$.

Now we are prepared to formulate a theorem in the oriented case which corresponds to Theorem 9.1. The creative first and main idea for this theorem is due to A.Guedes de Oliveira.

Theorem 9.3. *A map* $\chi : \Lambda(n,d) \to GF_3$ *is an oriented matroid if and only if* $\chi \in \mathcal{V}^O(n,d)) := \mathcal{V}(\mathcal{J}^O(n,d))$.

Definition 9.4. *We call* $\mathcal{V}^O(n,d)) := \mathcal{V}(\mathcal{J}^O(n,d))$ *the chirotope variety in rank d for n points.*

Once having introduced this algebraic language, it is very natural to ask for a "good" or "small" characterization of our chirotope variety. To stress this point, we pose it as a problem.

Problem 9.5. *Find a "good" set of polynomials for defining the chirotope variety* $\mathcal{V}^O(n,d)$.

It is the main object of this chapter to give a first answer to this problem. In this context, Theorem 9.3 will be seen as a key step. We also remark that a small set of polynomials for defining the chirotope variety might be useful in isomorphism tests for oriented matroids.

Proof of Theorem 9.3.
Given an oriented matroid, we have for any pair (μ, λ), $\mu \in \Lambda(n, d-1)$ and $\lambda \in \Lambda(n, d+1)$: the set of monomials $\{M_i(\chi)|1 \le i \le d+1\}$, defined as

$$\{(-1)^i\chi[\lambda_1,\ldots,\lambda_{i-1},\lambda_{i+1},\ldots,\lambda_{d+1}]\chi[\mu_1,\ldots,\mu_{d-1},\lambda_i]\},$$

is either equal to $\{0\}$, or it contains $\{-1,1\}$. For $P = od\{|\mu|\lambda\} \in \mathcal{J}^O(n,d)$, we have to show that

$$P(\chi) = \sum_{k\,odd} P^{(k)}_{\{|\mu|\lambda\}}(\chi) := \sum_{k\,odd} e^{(k)}(M_1(\chi),\ldots,M_{d+1}(\chi)) = 0.$$

Taking the last sum over the odd elementary symmetric functions twice, it can be rewritten as

$$(1 - M_1(\chi))\ldots(1 - M_{d+1}(\chi)) - (1 + M_1(\chi))\ldots(1 + M_{d+1}(\chi))$$

which is zero in the oriented matroid case.

On the other hand, given a map $\chi : \Lambda(n,d) \to GF_3$ with $\chi \in \mathcal{V}(\mathcal{J}^O(n,d))$, e.g. $od\{|\mu|\lambda\}(\chi) = 0$ for all pairs (μ, λ) as above, we have either $M_i(\chi) = 0$ for all i, $1 \le i \le d+1$, or there are two monomials $M_k(\chi), M_l(\chi)$ of opposite signs. $\square$

9.2. Syzygies for chirotope polynomials
In this section we provide some algebraic tools in order to apply them in Section 9.3. Whereas Grassmann-Plücker polynomials $\{|\mu|\lambda\}$ define the grassmannian, the *chirotope polynomials* $od\{|\mu|\lambda\}$ define the chirotope variety $\mathcal{V}^O(n,d)) := \mathcal{V}(\mathcal{J}^O(n,d))$. This close connection of both varieties is reflected when we study syzygies for the defining polynomials. Polynomial identities connecting sets of Grassmann-Plücker polynomials $\{|\mu|\lambda\}$ give rise to corresponding syzygies for the *chirotope polynomials* $od\{|\mu|\lambda\}$. This seems to cast a new "algebraic" light on the close connection between point configurations (represented on the grassmannian) and chirotopes (represented on the chirotope variety), and a further study of how these varieties are linked is suggested, see also [26].

We first collect some easy facts about our Grassmann-Plücker polynomials $\{|\mu|\lambda\}$ related to *chirotope polynomials* $od\{|\mu|\lambda\}$ in the following Lemma 9.6.

The symmetric group $S_n = S_E$, acts on the polynomial ring $GF_3[\Lambda(n,d)]$. The relabeling of the vertices according to $\sigma \in S_n = S_E$ will change the polynomial $P \in GF_3[\Lambda(n,d)]$ to $P^\sigma \in GF_3[\Lambda(n,d)]$. For $\lambda = [\lambda_1, \ldots, \lambda_{d+1}]$ and $\mu = [\mu_1, \ldots, \mu_{d-1}]$ with $\lambda_1 = \mu_1$, $\lambda_2 = \mu_2, \ldots$, $\lambda_k = \mu_k$, we also write

$$\{|\mu|\lambda\} =: \{\mu_1, \ldots, \mu_k | \mu_{k+1}, \ldots, \mu_{d-1} | \lambda_{k+1}, \ldots, \lambda_{d+1}\}.$$

We consider the maps od, or ev,
$od\{|\mu|\lambda\} := \sum_{k \, odd} P^{(k)}_{\{|\mu|\lambda\}}$, or $ev\{|\mu|\lambda\} := \sum_{k \, even} P^{(k)}_{\{|\mu|\lambda\}}$, also to be defined as maps for vectors $(P^{(k)}_{\{|\mu|\lambda\}})_{k \, odd}$, or $(P^{(k)}_{\{|\mu|\lambda\}})_{k \, even}$, respectively. The following lemma can be applied when we want to construct syzygies for chirotope polynomials. For instance, they were also useful in generating the formula presented later in Theorem 9.9.

Lemma 9.6.
(i) For any permutation $\sigma = (\sigma_1, \sigma_2) \in S_\mu \times S_\lambda \subset S_E$, we have

$$\{|\mu|\lambda\}^\sigma = (-1)^{\text{sign}(\sigma)} \{|\mu|\lambda\}.$$

(ii) With the above definition of the maps od and ev, we have

$$od(A, B) = od(A)ev(B) + od(B)ev(A),$$

$$ev(A, B) = od(A)od(B) + ev(B)ev(A),$$

$$ev(A)od(-B, C) + ev(B)od(-C, A) + ev(C)od(-A, B) = 0,$$

$$od(A)od(-B, C) + od(B)od(-C, A) + od(C)od(-A, B) = 0.$$

(iii) The following equations hold modulo GF_3 :

$$c \, od(a_1, \ldots, a_k) = od(c \, a_1, \ldots, c \, a_k),$$

$$c \, od\{\nu|\mu|\lambda\} = od(c \, \{\nu|\mu|\lambda\}).$$

(iv) For $P = (M_1, \ldots, M_k)$, we have:

$$2od(P) = \prod_i (1 + M_i) - \prod_i (1 - M_i),$$

$$2ev(P) = \prod_i (1 + M_i) + \prod_i (1 - M_i).$$

All proofs of Lemma 9.6 are straightforward, and we omit them here.

Our next Lemma 9.7. provides us with a particular syzygy. This algebraic dependence in the description of the grassmannian will be used later to prove a corresponding assertion for the chirotope variety.

Lemma 9.7. With the following abbreviations $P^0_{\mu,\lambda} := \{|\mu|\lambda\} = \{|\mu_1,\dots,\mu_{d-1}|\lambda_1,\lambda_2,\dots,\lambda_{d+1}\}$,

$$P^1_{\mu,\lambda} := \{\mu_1|\mu_2,\dots,\mu_{d-1}|\lambda_2,\lambda_3,\dots,\lambda_{d+1}\},$$
$$P^2_{\mu,\lambda} := \{\mu_1|\mu_2,\dots,\mu_{d-1}|\lambda_1,\lambda_3,\dots,\lambda_{d+1}\},$$
$$P^i_{\mu,\lambda} := \{\lambda_3,\dots,\lambda_{i-1},\lambda_{i+1},\dots,\lambda_{d+1}|\mu_1,\lambda_1,\lambda_2,\lambda_i\}, \quad \text{for } 3 \le i \le d+1,$$

we have the syzygy

$$S_{\mu,\lambda} := [\mu_1,\lambda_3,\dots,\lambda_{d+1}]\,P^0_{\mu,\lambda} - [\lambda_1,\lambda_3,\dots,\lambda_{d+1}]\,P^1_{\mu,\lambda}$$

$$+ [\lambda_2,\lambda_3,\dots,\lambda_{d+1}]\,P^2_{\mu,\lambda} + \sum_{i=3}^{d+1}[\mu_1,\dots,\mu_{d-1},\lambda_i]\,P^i_{\mu,\lambda} = 0.$$

Remark 9.8.
The proof of Lemma 9.7. proceeds by straightforward calculations, and although we omit it here, a special case is seen in the proof of Theorem 9.9. In general, it is more difficult to find appropriate syzygies then to show that they hold. This is the stage where computer algebra software can help, in particular the Gröbner basis technique can be applied in the case of the Galois field GF_3.

Whereas Lemma 9.7. describes a special dependence for polynomials that define the grassmannian, the next theorem provides us with a corresponding result for those polynomials defining our chirotope variety. In the following we adjust everything to the rank 3 case which already shows the possibilities of this method. We choose $d = 3$; $\mu = \mu_1,\mu_2$; $\lambda = \lambda_1,\dots,\lambda_4$.

Theorem 9.9. Let the polynomial $L_{\mu|\lambda} \in GF_3[\Lambda(n,3)]$ be defined as

$$L_{\mu|\lambda} := [\mu_1,\lambda_3,\lambda_4](1 - [\lambda_2,\lambda_3,\lambda_4]^2[\mu_1,\mu_2,\lambda_3][\mu_1,\mu_2,\lambda_4][\mu_1,\lambda_1,\lambda_3][\mu_1,\lambda_1,\lambda_4]).$$

Then we have

$$L_{\mu|\lambda}\,od(P^0_{\mu,\lambda}) \in \; < \{od(P^i_{\mu,\lambda})|1 \le i \le 4\} > .$$

Remark 9.10. Note that a permutation of the vertices changes $od(P^0_{\mu,\lambda})$ at most up to a sign whereas $L_{\mu|\lambda}$ is changed in general. It will later be shown that by permuting the elements of μ and λ, for any χ and for any $od(P^0_{\mu,\lambda})(\chi)$ there is a particular polynomial $L_{\mu|\lambda}(\chi)$ different from zero. This implies that in deleting $od(P^0_{\mu,\lambda})$ in the definition of the chirotope variety, we still get the same set of chirotopes.

Proof of Theorem 9.9. In principle, the proof can be done automatically, provided a good computer algebra package is available. Computing a Gröbner-basis for the ideal $< \{od(P^i_{\mu,\lambda})|1 \le i \le 4\} >$ and reducing the polynomial

$L_{\mu,\lambda}od(P^0_{\mu,\lambda})$ modulo this ideal would give the answer. The polynomial lies in the ideal if and only if it can be reduced to zero, see e.g. [40]. We tried this approach with MACSYMA on a VAX/VMS-8500 computer but we did not succeed in finding a Gröbner-basis within less than 1 CPU-hour. There is a more mathematical approach to describe how the syzygy was found, and indeed the proof can be done in this way. This would mean applying Lemma 9.6., and in order to get a small polynomial $L_{\mu,\lambda}$, this has to be done with great care when handling boring polynomials.

In order to achieve a proof of the theorem, we prefer choosing a way in between, namely by reducing the problem first until it can be solved by MACSYMA (or any other computer algebra package like MACAULAY, MAPLE, REDUCE, SACII) without difficulties. This will also tell us how Lemma 9.7. is involved.

In order to do so, it is convenient to have the polynomials $P^i_{\mu,\lambda}$ in our special case (rank 3, $\mu_1 = 1, \mu_2 = 2, \lambda_1 = 3, \lambda_2 = 4, \lambda_3 = 5, \lambda_4 = 6,$) explicitly:

$$P^0_{12,3456} = -[4,5,6][1,2,3] + [3,5,6][1,2,4] - [3,4,6][1,2,5] + [3,4,5][1,2,6]$$

$$P^1_{12,3456} = [1,5,6][1,2,4] - [1,4,6][1,2,5] + [1,4,5][1,2,6]$$

$$P^2_{12,3456} = [1,5,6][1,2,3] - [1,3,6][1,2,5] + [1,3,5][1,2,6]$$

$$P^3_{12,3456} = [6,1,3][6,4,5] - [6,1,4][6,3,5] + [6,1,5][6,3,4]$$

$$P^4_{12,3456} = [5,1,3][5,4,6] - [5,1,4][5,3,6] + [5,1,6][5,3,4]$$

Lemma 9.7 in this special case reads as follows (brackets were sorted already):

$$+[1,5,6] * P^0_{12,3456} =$$
$$-[1,2,3][1,5,6][4,5,6] + [1,2,4][1,5,6][3,5,6]$$
$$-[1,2,5][1,5,6][3,4,6] + [1,2,6][1,5,6][3,4,5]$$

$$+[4,5,6] * P^2_{12,3456} =$$
$$+[1,2,3][1,5,6][4,5,6] - [1,2,5][1,3,6][4,5,6] + [1,2,6][1,3,5][4,5,6]$$

$$-[3,5,6] * P^1_{12,3456} =$$
$$-[1,2,4][1,5,6][3,5,6] + [1,2,5][1,4,6][3,5,6] - [1,2,6][1,4,5][3,5,6]$$

$$+[1,2,5] * P^3_{12,3456} =$$
$$+[1,2,5][1,5,6][3,4,6] + [1,2,5][1,3,6][4,5,6] - [1,2,5][1,4,6][3,5,6]$$

$$+[1,2,6] * P^4_{12,3456} =$$
$$-[1,2,6][1,5,6][3,4,5] - [1,2,6][1,3,5][4,5,6] + [1,2,6][1,4,5][3,5,6]$$

Note that only 13 brackets are involved:

$$[1,2,3] \quad [1,2,4] \quad [1,2,5] \quad [1,2,6] \quad [1,3,5] \quad [1,3,6] \quad [1,4,5]$$

$$[1,4,6] \quad [1,5,6] \quad [3,4,5] \quad [3,4,6] \quad [3,5,6] \quad [4,5,6].$$

Moreover, we can reduce the number of variables down to 8. For the monomials in the above expressions, we use the new variables $A, B, C, \ldots, H$ as follows (compare the order above):

$$+[1,5,6] * P^0_{12,3456} = -A + B - C + D,$$
$$+[4,5,6] * P^2_{12,3456} = +A - E + F,$$
$$-[3,5,6] * P^1_{12,3456} = -B + G - H,$$
$$+[1,2,5] * P^3_{12,3456} = +C + E - G,$$
$$+[1,2,6] * P^4_{12,3456} = -D - F + H.$$

Thus we have

$$[1,5,6] * od(P^0_{12,3456}) = -A + B - C + D + ABC - ABD + ACD - BCD,$$
$$+[4,5,6] * od(P^2_{12,3456}) = +A - E + F - AEF,$$
$$-[3,5,6] * od(P^1_{12,3456}) = -B + G - H + BGH,$$
$$+[1,2,5] * od(P^3_{12,3456}) = +C + E - G - CEG,$$
$$+[1,2,6] * od(P^4_{12,3456}) = -D - F + H + DFH,$$

and it suffices to show that the polynomial $(1 - EF) * [1,5,6] * od(P^0_{12,3456})$, e.g. a multiple of the first polynomial in the variables $A, B, \ldots, H$, lies in the ideal generated by the last four polynomials. This can now be done in using any computer algebra system with the Gröbner-bases package. We add a shortened version of a corresponding session.

This is Macsyma 412.61 for DEC VAX 8500 Series Computers.
Copyright (c) 1982 Massachusetts Institute of Technology.
All Rights Reserved.
Enhancements (c) 1982, 1988 Symbolics, Inc. All Rights Reserved.
(D1) $-AEF + F - E + A$
(D2) $+BGH - H + G - B$
(D3) $-CEG - G + E + C$
(D4) $+DFH + H - F - D$
(D5) $A^3 - A$
(D6) $B^3 - B$
etc.
(D12) $H^3 - H$
(C13) grobner([d1,d2,d3,d4,d5,d6,d7,d8,d9,d10,d11,d12]);
(computes a Gröbner basis for the above polynomials)
(with respect to the lexicographical order)

(D13)/R/ $[(((-F+E)B+AF-AE)C+(-AF+AE)B+F-E)D$
$+((AF-AE)B-F+E)C+(F-E)B-AF+AE,$
$((-B-F)G+FB+1)D+(-FB-1)G+B+F,$
$(-H+F)D+(-F^2+1)H^2-FH+F^2,(F^2-1)D+(-F^2+1)H,$
$((-A^2+1)E^2+A^2-1)D+((A^2-1)E^2-A^2+1)H,$
$((-B+E)H+EB-1)C+(-EB+1)H+B-E,$
$(G-E)C+(E^2-1)G^2+EG-E^2,(-E^2+1)C+(E^2-1)G,$
$(-G+B)H+(-B^2+1)G^2-BG+B^2,(B^2-1)H+(-B^2+1)G,$
$(E-A)F+(A^2-1)E^2+AE-A^2,(-A^2+1)F+(A^2-1)E,A^3-A,B^3-B,$
$C^3-C,D^3-D,E^3-E,F^3-F,G^3-G,H^3-H]$
(D14)$-BCD+ACD-ABD+D+ABC-C+B-A$
(D15) $(-BCD+ACD-ABD+D+ABC-C+B-A)(1-EF)$
(C16) id-member(d15,d13);
(returns TRUE if and only if polynomial (d15) is a member of the ideal (d13).)
; Starting garbage collection due to dynamic-0 space overflow.
; Finished garbage collection due to dynamic-0 space overflow.
(D16) TRUE   (which proves our theorem!)
(C17) id-member(d14,d13);
; Starting garbage collection due to dynamic-1 space overflow.
; Finished garbage collection due to dynamic-1 space overflow.
(D17) FALSE

The last result (D17) shows in addition that the factor $(1-EF)$ in the polynomial is essential. The proof does not show how the polynomial in question can be generated, but this is not needed in the conclusion we will make in the next section.

□

9.3. Applications for planar problems

Let us start this section with two very natural problems:

(1) Find a fast algorithm for testing whether a given map $\chi : \Lambda(n,d) \to GF_3$ is an oriented matroid.

(2) Find a fast algorithm for testing whether two oriented matroids are isomorphic up to relabeling the vertices and up to sign reversal.

We do not aim to provide a satisfactory answer for these very general questions, but we can see that finding a small set of conditions, which characterizes oriented matroids, is a useful and essential first step for both questions.

For the sake of simplicity, we deal in this section only with rank 3, the first interesting case. For oriented matroids in rank 3, we have both 3-term Grassmann-Plücker syzygies and 4-term Grassmann-Plücker syzygies. Both were used in our Definition 1.12 of partial chirotopes and of oriented matroids. It is our aim to give a new proof that shows that the 4-term Grassmann-Plücker syzygies are not needed in the definition of oriented matroids when we already know that the alternating function $\chi : \Lambda(n,3) \to GF_3$ is *matroidal*, e.g. if the set $\chi^{-1}(-1) \cup \chi^{-1}(1)$ forms the set of bases of a matroid. This is trivially true in case $\chi(\lambda) \neq 0$ for all $\lambda \in \Lambda(n,3)$.

Apart from the ideal $\mathcal{J}^O(n,3)$ generated by all polynomials $od\{|\mu|\lambda\}$, we consider the ideal $\mathcal{J}_3^O(n,3)$ generated only by the subset $\{ od\{|\mu|\lambda\} \mid \mu_1 = \lambda_1 \}$ of polynomials corresponding to 3-term syzygies. With $\mathcal{V}^O(n,3)) := \mathcal{V}(\mathcal{J}^O(n,3))$ and $\mathcal{V}_3^O(n,3)) := \mathcal{V}(\mathcal{J}_3^O(n,3))$, we have the corresponding varieties $\mathcal{V}^O(n,3)) \subset \mathcal{V}_3^O(n,3))$.

Compared to Theorem 9.3., we now state in Theorem 9.11. a sharper version of a characterization of oriented matroids (chirotopes). In the classical language of oriented matroids, a first proof for this assertion was given by Las Vergnas [102]. To show the equivalence of Las Vergnas's result with our Theorem 9.3, we need the Lawrence equivalence of oriented matroids and chirotopes, see [64], [59]. Remember also that Las Vergnas proved one direction of this equivalence earlier [102], [102]. And remember that Dress later independently gave another proof for this equivalence. Here it is the method of proof we are interested in. We achieve the result in a straightforward manner.

Theorem 9.11. *For any matroidal map* $\chi : \Lambda(n,3) \to GF_3$, *we have:* χ *is an oriented matroid if and only if* $\chi \in \mathcal{V}_3^O(n,3) := \mathcal{V}(\mathcal{J}_3^O(n,3))$.

Proof The proof is based on Theorem 9.9. There is only one non-trivial direction. We only have to show that for every matroidal map $\chi : \Lambda(n,3) \to GF_3$ with $\chi \in \mathcal{V}(\mathcal{J}_3^O(n,3))$ and for any polynomial $od(P_{\mu,\lambda}^0)$, we have $od(P_{\mu,\lambda}^0)(\chi) = 0$. In order to do this, we assume $od(P_{\mu,\lambda}^0)(\chi) \neq 0$, and in constructing an appropriate permutation for the given map χ such that $L_{\mu|\lambda}(\chi)$ is non-zero, we arrive at a contradiction to Theorem 9.9.

Since $od(P_{\mu,\lambda}^0)(\chi) \neq 0$, we can find a monomial $\chi(\ldots)\chi(\ldots) \neq 0$ and by relabeling the vertices if neccessary, we can assume $\chi[\mu_1, \mu_2, \lambda_1]\chi[\lambda_2, \lambda_3, \lambda_4] \neq 0$. Steinitz' exchange lemma tells us that we have more (i,j,k) with $\chi[i,j,k] \neq 0$, and we are still free in choosing first any permutation $\sigma \in S_2 \times S_4$ before we consider the polynomial

$$L_{\mu|\lambda} := [\mu_1, \lambda_3, \lambda_4](1 - [\lambda_2, \lambda_3, \lambda_4]^2[\mu_1, \mu_2, \lambda_3][\mu_1, \mu_2, \lambda_4][\mu_1, \lambda_1, \lambda_3][\mu_1, \lambda_1, \lambda_4]).$$

These ideas are used in the following. We write our polynomial $L_{\mu|\lambda}$ also in the form $L_{\mu|\lambda} = X_{\mu|\lambda}(1 - Y_{\mu|\lambda})$ with $X_{\mu|\lambda} := [\mu_1, \lambda_3, \lambda_4]$ and $Y_{\mu|\lambda} := [\lambda_2, \lambda_3, \lambda_4]^2[\mu_1, \mu_2, \lambda_3][\mu_1, \mu_2, \lambda_4][\mu_1, \lambda_1, \lambda_3][\mu_1, \lambda_1, \lambda_4]$ In denoting the three elements of the cyclic group generated by $(\lambda_1, \lambda_3, \lambda_4)$ with $\sigma_1, \sigma_2, \sigma_3$, we see

$$\sigma_1(Y_{\mu|\lambda})(\chi)\sigma_2(Y_{\mu|\lambda})(\chi)\sigma_3(Y_{\mu|\lambda})(\chi) = (\chi[\ldots])^2 \ldots (\chi[\ldots])^2 \times$$

$$\times \chi[\mu_1, \lambda_1, \lambda_3]\chi[\mu_1, \lambda_1, \lambda_4]\chi[\mu_1, \lambda_3, \lambda_4]\chi[\mu_1, \lambda_3, \lambda_1]\chi[\mu_1, \lambda_4, \lambda_1]\chi[\mu_1, \lambda_4, \lambda_3] \neq 1,$$

which in turn tells us that $1 - \sigma_i(Y_{\mu|\lambda})(\chi)$ must be different from zero for at least one i, $1 \leq i \leq 3$.

The same holds true when we replace μ_1 with μ_2 and vice versa. Next we observe that in starting with $\chi[\mu_1, \mu_2, \lambda_1] \neq 0$ and $\chi[\lambda_2, \lambda_3, \lambda_4] \neq 0$ and applying Steinitz' exchange lemma twice, we can find at least one additional basis among the following corresponding brackets

$$[\mu_1, \lambda_3, \lambda_4], [\mu_1, \lambda_1, \lambda_3], [\mu_1, \lambda_4, \lambda_1], [\mu_2, \lambda_3, \lambda_4], [\mu_2, \lambda_1, \lambda_3], [\mu_2, \lambda_4, \lambda_1].$$

We assume w.l.o.g $\chi(\mu_1, \lambda_3, \lambda_4) \neq 0$, otherwise exchange μ_1 with μ_2 and/or relabel the vertices $\lambda_1, \lambda_3, \lambda_4$. If $\chi[\mu_1, \lambda_1, \lambda_3] = 0$ or $\chi[\mu_1, \lambda_1, \lambda_4] = 0$, the second factor in $L_{\mu|\lambda}$ is 1, and we have nothing more to prove. But otherwise, we find a suitable element σ_i with $\sigma_i(L_{\mu|\lambda})(\chi) \neq 0$ with the argument used above. $\square$

For more results about this algebraic approach and for a corresponding proof of Theorem 9.11 in case of arbitrary rank d, we refer again to [26].

References

[1] N. Alon : "The number of polytopes, configurations and real matroids". *Mathematika* **33** (1986) 62–71.

[2] A. Altshuler : "Neighborly combinatorial 3-manifolds with 9 vertices. *Discrete Math.* **8** (1974) 113–137.

[3] A. Altshuler : "Neighborly 4-polytopes and neighborly combinatorial 3-manifolds with ten vertices". *Canadian J. Math.* **29** (1977) 400–420.

[4] A. Altshuler : "A remark on the politopality of an interesting 3-sphere". *Israel Journal of Mathematics* **48** (1984), 159–160.

[5] A. Altshuler, J. Bokowski, L. Steinberg : "The classification of simplicial 3-spheres with nine vertices into polytopes and non-polytopes". *Discrete Math.* **31** (1980) 115–124.

[6] A. Altshuler, I. Shemer : "Construction theorems for polytopes". *Israel Journ. Math.* **47** (1984) 99–110.

[7] C. Antonin: "Ein Algorithmusansatz für Realisierungsfragen im E^d getestet an kombinatorischen 3-Sphären". Staatsexamensarbeit, Bochum 1982.

[8] M.F. Atiyah, I.G. MacDonald : "Introduction to Commutative Algebra", Addison-Wesley Publ., Reading, Massachusetts, 1969.

[9] A. Bachem : "Convexity and Optimization in Discrete Structures"; in P. Gruber, J. Wills (ed.): *Convexity and Applications*, Birkhäuser, Basel, 1983.

[10] D. Bayer, M. Stillman : "Computational Algebraic Geometry". Short course presented at *Computers and Mathematics*, Stanford University, July 1986.

[11] M. Bayer, B. Sturmfels : "Lawrence polytopes", *Canadian J. Mathematics*, to appear.

[12] E. Becker : "On the real spectrum of a ring and its applications to semi-algebraic geometry". *Bulletin Amer. Math. Soc.* **15** (1986) 19–60.

[13] M. Ben-Or, D. Kozen, J. Reif : "The complexity of elementary algebra and geometry", in "Theory of Computing", Proc. of the 16th Annual ACM Symposium (Wash. DC, April 1984), A.C.M., New York, 1984.

[14] W. Biena, I.P. da Silva : "On the inversion of the sign of one base in an oriented matroid". Preprint, University Paris 6, 1986.

[15] L.J. Billera, B.S. Munson : "Polarity and inner products in oriented matroids". *European J. Combinatorics* **5** (1984) 293–308.

[16] L.J. Billera, B.S. Munson: "Triangulations of oriented matroids and convex polytopes". *SIAM J. Algebraic Discrete Meth.* **5** (1984) 515–525.

[17] R. Bland, M. Las Vergnas : "Orientability of matroids". *J. Combinatorial Theory* **B 24** (1978) 94–123.

[18] J. Bokowski : "A geometric realization without self-intersections does exist for Dyck's regular map". *Discrete Comput. Geometry*, to appear.

[19] J. Bokowski :"Aspects of computational synthetic geometry, II: Combinatorial complexes and their geometric realization - an algorithmic approach". *Proceedings of Computer-aided Geometric Reasoning, INRIA, Antibes (France)*, 1987.

[20] J. Bokowski, U. Brehm : "A new polyhedron of genus 3 with 10 vertices. *Colloquia Math. Soc. János Bolyai, Siofok, 1985.*

[21] J. Bokowski, U. Brehm : "A polyhedron of genus 4 with minimal number of vertices and maximal symmetry". *Geometriae Dedicata* **29** (1989) 53–64.

[22] J. Bokowski, A. Eggert : "All realizations of Möbius' torus with 7 vertices". Preprint 1009, TH Darmstadt, 1986.

[23] J. Bokowski, G. Ewald, P. Kleinschmidt : "On combinatorial and affine automorphisms of polytopes". *Israel Journ. Math.* **47** (1984) 123–130.

[24] J. Bokowski, K. Garms : "Altshuler's sphere M_{425}^{10} is not polytopal". *European J. Combinatorics* **8** (1987) 227–229.

[25] J. Bokowski, A. Guedes de Oliveira: "Simplicial convex 4-polytopes do not have the isotopy property". *Portugaliae Mathematica* , to appear.

[26] J. Bokowski, A. Guedes de Oliveira, J. Richter: "Matroids and oriented matroids as algebraic varieties". Manuscript 1988.

[27] J. Bokowski, J. Richter: "On the finding of final polynomials", *Europ. J. Comb.*, to appear.

[28] J. Bokowski, J. Richter: "Bi-quadratic final polynomials and embedded symmetric polyhedra", Manuscript 1988.

[28a] J. Bokowski, J. Richter, B. Sturmfels: "Nonrealizability proofs in computational geometry". *Discrete Comput. Geometry* to appear.

[29] J. Bokowski, I. Shemer : "Neighborly 6-polytopes with 10 vertices". *Israel Journ. Math.* **58** (1987) 103–124.

[30] J. Bokowski, B. Sturmfels : "An infinite family of minorminimal nonrealizable 3-chirotopes", *Math. Z.* **200** (1989) 583–589.

[31] J. Bokowski, B. Sturmfels : "On the coordinatization of oriented matroids". *Discrete Comput. Geometry* **1** (1986) 293–306.

[32] J. Bokowski, B. Sturmfels : "Polytopal and non-polytopal spheres - An algorithmic approach", *Israel Journ. Math.* **57** (1987) 257–271.

[33] J. Bokowski, B. Sturmfels : "Reell realisierbare orientierte Matroide". *Bayreuther Math. Schriften* **21** (1986) 1–13.

[34] J. Bokowski, B. Sturmfels : "Programmsystem zur Realisierung orientierter Matroide". *Preprint Cologne* **21** (1986) 1–13.

[35] J. Bokowski, J. Wills : "Regular polyhedra with hidden symmetries", *The Mathematical Intelligencer* **10** No.1 (1988) 27–32.

[36] K.-H. Borgwardt : "Some distribution-independent results about the asymptotic order of the average number of pivot steps of the simplex method". *Math. of Operations Research* **7** (1982) 441–462.

[37] E. Boros, Z. Füredi, L.M. Kelly : On representing Sylvester-Gallai designs. *Discrete Comput. Geometry*, in print.

[38] N. Bourbaki : "Éléments de Mathematique, Algébre, Ch. 1 à 3". Hermann, Paris, 1970.

[39] U. Brehm: "Maximally symmetric realizations of Dyck's regular map". *Mathematika* **34** (1987) 229–236.

160

[40] B. Buchberger : "Gröbner Bases - an Algorithmic Method in Polynomial Ideal Theory", Chapter 6 in N.K. Bose (ed.) : Multidimensional Systems Theory, D. Reidel, 1985.

[40a] J.R. Buchi, W.E. Fenton: "Large convex sets in oriented matroids", *J. Combinatorial Theory B* **45** (1988) 293–304.

[41] C. Buchta : "Zufällige Polyeder - eine Übersicht", in E. Hlawka (ed.): *Zahlentheoretische Analysis*, Wiener Seminarberichte 1980-82, Lecture Notes in Mathematics **1114**, 1–13, Springer, Heidelberg, 1985.

[42] J. Canny : "Some algebraic and geometric computations in PSPACE", Proc. 20th ACM Symposium on the Theory of Computing, Chicago, May 1988.

[43] C. Collins : "Quantifier Elimination for Real Closed Fields by Cyclindrical Algebraic Decomposition", in H. Brakhage (ed.): *Automata Theory and Formal Languages*, Lecture Notes in Computer Science **33**, 134–163, Springer, Heidelberg, 1975.

[44] R. Cordovil : "Oriented matroids of rank three and arrangements of pseudolines", *Annals of Discrete Math.* **17** (1983) 219–223.

[45] R. Cordovil, P. Duchet : "On the number of sign invariant pairs of points in oriented matroids". *Discrete Math.*, to appear.

[46] R. Cordovil, P. Duchet : "Oriented matroids and cyclic polytopes". *Combinatorica*, to appear.

[47] J. van de Craats : "On Simonis' 10_3 configuration". *Nieuw Archief voor Wiskunde* **4** (1983) 193–207.

[48] H. Crapo, J. Ryan : "Spatial realizations of linear scenes", *Structural topology* **13** (1986) 33–68.

[49] G. Crippen, T. Havel : "Distance Geometry and Molecular Conformation", Chemometrics Research Studies Press, Letchworth, U.K., 1988.

[50] A. Császár : "A polyhedron without diagonals", *Acta Sci. Math. Szeged.* **13** (1949) 140–142.

[51] R. Daublebsky von Sterneck : "Die Configurationen 11_3". *Monatshefte für Mathematik und Physik* **5** (1894) 325–330.

[52] R. Daublebsky von Sterneck : "Die Configurationen 12_3". *Monatshefte für Mathematik und Physik* **6** (1895) 223–255.

[53] J. Dieudonné : "The tragedy of Grassmann". *Linear and Multilinear Algebra* **8** (1979) 1–14.

[54] J. Dieudonné, J.B. Carell : "Invariant theory, old and new", Academic Press, New York, 1971.

[55] P. Doubilet, J.-C. Rota, J. Stein : "On the foundations of combinatorial theory : IX. Combinatorial methods in invariant theory", *Studies in Appl. Math.* **LIII**, No. 3 (1974) 185–216.

[56] A. Dreiding, K. Wirth : "The multiplex - A classification of finite ordered point sets in oriented d-dimensional space". *Math. Chemistry* **8** (1980) 341–352.

[57] A. Dress : "Chirotopes and Oriented Matroids". *Bayreuther Math. Schriften* **21** (1986) 14–68.

[58] A. Dress : "Duality theory for finite and infinite matroids with coefficients". *Advances in Math.* **59** (1986) 97–123.

[59] A. Dress, W. Wenzel : "Grassmann-Plücker relations and matroids with co-efficients" *Advances in Mathematics* , submitted.

[60] D.W. Dubois : "A nullstellensatz for ordered fields". *Arkiv för Matematik* **8** (1969) 111-114.

[61] R.A. Duke : "Geometric embedding of complexes", *American Math. Monthly* **77** (1970) 597-603.

[62] G. Ewald, P. Kleinschmidt, U. Pachner, C. Schulz : "Neuere Entwicklungen in der kombinatorischen Konvexgeometrie", in J.Tölke, J.M.Wills (ed.) : *Contributions to geometry*, Proc. of the Geometry-Symp., Siegen, Birkhäuser, Basel, 1978.

[63] N.E. Fenton : "Matroid representations - an algebraic treatment", *Quart. J. Math. Oxford, Ser.* (2) **35** (1984) 263-280.

[64] J. Folkman, J. Lawrence : "Oriented Matroids". *J. Combinatorial Theory B* **25** (1978) 199-236.

[65] K. Fukuda: "Oriented matroid programming". Ph.D. Thesis, University of Waterloo, 1982.

[66] D. Gale : "Neighboring vertices on a convex polyhedron". In H.W. Kuhn, A.W. Tucker (ed.) : Linear inequalities and related systems, Princeton University Press, Princeton, 1956.

[67] D. Garbe:"Über die regulären Zerlegungen geschlossener Flächen". *J. Reine Angew. Math.*, **237** (1969) 39-55.

[68] I.M. Gelfand, R.M. Goresky, R.D. MacPherson, V. Serganova : "Combinatorial geometries, convex polyhedra and Schubert cells". *Advances in Math.*, **63** (1987) 301-316.

[69] J.E. Goodman, R. Pollack : "Proof of Grünbaum's conjecture on the stretchability of certain arrangements of pseudolines". *J. Combinatorial Theory A* **29** (1980) 385-390.

[70] J.E. Goodman, R. Pollack : "Multidimensional Sorting". *SIAM J. Comput.* **12** (1983) 484-507.

[71] J.E. Goodman, R. Pollack : "Upper bounds for configurations and polytopes in R^d". *Discrete Comput. Geometry* 1 (1986) 219-227.

[72] J.E. Goodman, R. Pollack, in *Tagungsbericht Oberwolfach*, Tagung über kombinatorische Geometrie, September 1984, List of problems.

[73] J. Gray: "From the history of a simple group". *The Mathematical Intelligencer* 4 (1982), 59-66.

[74] W.H. Greub : "Multilinear Algebra". Springer, Berlin, 1967.

[75] D.Yu. Grigor'ev, N.N. Vorobjev : "Solving systems of polynomial inequalities in subexponential time". *J. Symbolic Computation* 5 (1988) 37-64.

[76] B. Grünbaum : "Convex Polytopes". Interscience Publ., London, 1967.

[77] B. Grünbaum: "Arrangements and spreads". American Math. Soc., Regional Conf. Ser. 10 (1972).

[78] B. Grünbaum : "Notes on configurations". Lectures presented in the *Combinatorics and Geometry Seminar*, Univ. of Washington, 1986.

[79] B. Grünbaum : "The importance of being straight". Proc. 12th Bienn. Intern. Seminar of the Canad. Math. Congress, Vancouver 1969 (1970), 243-254.

[80] B. Grünbaum : "Arrangements of hyperplanes". Proc. 2nd Louisiana Conf. on Combinatorics, Graph Theory and Computing. Louisiana State University, Baton Rouge (1971), 41–106.

[81] B. Grünbaum : "The real configuration 21_4". Manuscript, University of Washington, 1986.

[82] B. Grünbaum, V.P. Sreedharan : "An enumeration of simplicial 4-polytopes with 8 vertices". J. Combinatorial Theory 2 (1967) 437–465.

[83] G.H. Hardy, E.M. Wright : "An Introduction to the Theory of Numbers". Clarendon Press, Oxford, 1979.

[84] R. Hartshorne : "Algebraic Geometry". Springer, New York, 1977.

[85] D. Hilbert, S. Cohn-Vossen : "Geometry and the Imagination", (English Edition), Chelsea, New York, 1983.

[86] W.V.D. Hodge, D. Pedoe, : "Methods of Algebraic Geometry, Volume I", Cambridge University Press, 1947.

[87] J.F.P. Hudson : "Piecewise Linear Topology", W.A. Benjamin, New York, 1969.

[88] N. Jacobson : "Lectures in Abstract Algebra", Volume III, v. Norstrand, Princeton, 1964.

[89] B. Jaggi, P. Mani-Levitska, B. Sturmfels, N. White : "Uniform oriented matroids without the isotopy property", Discrete Comput. Geometry 4 (1989) 97–100.

[90] B. Jaggi, P. Mani-Levitska : "A simple arrangement of lines without the isotopy property", Manuscript, Bern, Switzerland, January 1988.

[91] G. Kalai : "Many triangulated spheres", Discrete Comput. Geometry 3 (1988) 1–14.

[92] L.M. Kelly : "A resolution of the Sylvester-Gallai problem of J.P. Serre", Disrete Comput. Geometry 1 (1986) 101–104.

[93] V. Klee, P. Kleinschmidt : "The d-step conjecture and its relatives". Math. of Operations Research, 12 (1987) 718–755.

[94] V. Klee, S. Wagon : "Unsolved Problems in Mathematics", in preparation.

[95] J.P.S. Kung : "Bimatroids and invariants". Advances in Math. 30 (1978) 238–249.

[96] J.P.S. Kung : "A Source Book in Matroid Theory", Birkhäuser, Boston, 1986.

[97] J.P.S. Kung : "Combinatorial geometries representable over $GF(3)$ and $GF(q)$. I. The number of points". Preprint, North Texas State University, Denton, 1987.

[98] E. Kunz : "Introduction to commutative algebra and algebraic geometry", Birkhäuser, Boston, 1985.

[99] B. Kutzler, S. Stifter : "On the application of Buchberger's algorithm to automated geometry theorem proving", J. Symbolic Computation 2 (1986) 389–397.

[99a] M. Las Vergnas : "Bases in oriented matroids". J. Combinatorial Theory B 25 (1978) 283–289.

[100]. M. Las Vergnas : "Convexity in oriented matroids". J. Combinatorial Theory B 29 (1980) 231–243.

[101] M. Las Vergnas : "Order properties of lines in the plane and a conjecture of G.Ringel". *J. Combinatorial Theory* **B 41** (1986) 246–249.

[102] M. Las Vergnas : "Matroides orientables". *C.R. Acad. Sci. Paris* **280** (1975) 61–64.

[103] R. Lauffer : "Die nichtkonstruierbare Konfiguration (10_3)". *Mathematische Nachrichten* **11** (1954) 303–304.

[104] J. Lawrence : "Oriented matroids and multiply ordered sets". *Linear Algebra and its Appl.* **48** (1982) 1–12.

[105] H.R. Lewis, C.H. Papadimitriou : "Elements of the Theory of Computation". Prentice Hall, Englewood Cliffs, N.J., 1981.

[106] B. Lindström : "On the realization of convex polytopes, Euler's formula and Möbius functions". *Aequationes mathematicae* **6** (1971) 235–239.

[107] D. Ljubic : "A new torus with quadrangular faces". Presented at the A.M.S. Regional Conference, Tacoma, Washington, June 1987.

[108] D. Ljubic, J.P. Roudneff, B. Sturmfels : "Arrangements of lines and pseudolines without adjacent triangles". *J. Combinatorial Theory* **A 50** (1989) 24–32.

[109] A. Mandel : "Topology of Oriented Matroids". Ph.D. Thesis, University of Waterloo, 1982.

[110] P. Mani : "Automorphismen von polyedrischen Graphen". *Math. Ann.* **192** (1971) 279–303.

[111] P. Mani : Letter to V. Klee from April 1987.

[112] Y.V. Matiyasevic, "Diophantine representation of enumerable predicates", *Izvestia Akad. Nauk SSSR* **35** (1971), 3–30.

[113] B. Mazur, "Arithmetic on curves", *Bulletin Amer. Math. Soc.* **14** (1986) 207–259.

[114] S. MacLane : "Some interpretations of abstract linear dependence in terms of projective geometry", *American J. Math.* **58** (1936) 236–240.

[115] P. McMullen : "Transforms, Diagrams and Representations". In J. Tölke, J.M. Wills (ed.): *"Contributions to Geometry"*. Proc. of the Geometry-Symp. Siegen, June 1978, Birkhäuser Basel.

[116] P. McMullen, G.C. Shephard : "Convex Polytopes and the Upper Bound Conjecture". London Mathematical Society Lecture Note Series 3, Cambridge University Press, 1971.

[117] P. McMullen, E. Schulte, and J.M. Wills, "Infinite series of combinatorially regular polyhedra in three-space. *Geometriae Dedicata*, in print.

[118] P. McMullen, C. Schulz, and J.M. Wills, "Polyhedral 2-manifolds in E^3 with unusually large genus. *Israel J. Math.*, **46** (1983), 127–144.

[119] N.E. Mněv : "On the realizability over fields of the combinatorial type of convex polytopes" (in Russian). *Zap. Nauchn. Sem. Leningr. Otdel. Mat. Inst. Steklov* **123** (1983) 203–207.

[120] N.E. Mněv : "On manifolds of combinatorial types of projective configurations and convex polyhedra". *Soviet Math. Dokl.* **32** No.1 (1985) 335–337.

[121] N.E. Mnëv : "The universality theorems on the classification problem of configuration varieties and convex polytopes varieties", in O.Y. Viro (ed.): *Topology and Geometry – Rohlin Seminar*, Lecture Notes in Mathematics **1346**, 527–544, Springer, Heidelberg, 1988.

[121a] L.G. Novoa :"On n-ordered sets and order completeness". *Pacific J. Math.* **15** No.4 (1965) 1337–1345.

[122] L. Nachbin : "Integral de Haar", Instituto de Fisica e Matemática, Universidade do Recife, Brazil, 1960.

[123] F.P. Preparata, M.I. Shamos : "Computational Geometry", Springer, New York, 1985.

[124] J. Richter, B. Sturmfels : "On the topology and geometric construction of oriented matroids and convex polytopes", *Trans. Amer. Math. Soc.*, in print.

[125] J. Richter : "Kombinatorische Realisierbarkeitskriterien für orientierte Matroide", Diplomarbeit, Technische Hochschule Darmstadt, 1988.

[126] G. Ringel : "Teilungen der Ebene durch Geraden oder topologische Geraden". *Math. Zeitschrift* **64** (1956) 79–102.

[127] G. Ringel : "Über Geraden in allgemeiner Lage". *Elemente d. Math.* **12** (1957) 75–82.

[128] J.-P. Roudneff : "On the number of triangles in simple arrangements of pseudolincs in the real projective plane". *Discrete Mathematics* **60** (1986), 243–251.

[129] J.-P. Roudneff : "Matroides orientés et arrangements de pseudodroites". Thése de 3em cycle, L'Université Pierre et Marie Curie, Paris, 1986.

[130] J.-P. Roudneff : "Inseparability graphs of oriented matroids". Manuscript, Paris, 1987.

[131] J.-P. Roudneff, B. Sturmfels : "Simplicial cells in arrangements and mutations of oriented matroids". *Geometriae Dedicata* **27** (1988) 153–170.

[132] W. Rudin : "Functional Analysis". MacGraw-Hill, New York, 1973.

[133] L.A. Santalo : "Introduction to Integral Geometry", Hermann, Paris, 1953.

[134] R. Schneider : "Integralgeometrie". Vorlesungsskipt, Universität Freiburg, 1981.

[135] H. Schröter : "Über die Bildungsweise und geometrische Construction der Configurationen 10_3". *Nachrichten Gesellschaft der Wissenschaften Göttingen* (1889) 193–236.

[136] E. Schulte, J.M. Wills: "A polyhedral realization of Felix Klein's map $\{3,7\}_8$ on a Riemann surface of genus 3, *J. London Math. Soc.* 2(32) (1986), 253–262.

[137] J.T. Schwartz, M. Sharir : "On the "piano movers" problem. I. The case of a two-dimensional rigid polygonal body moving amidst polygonal barriers", *Comm. on Pure Appl. Math.* **XXXVI** (1983) 345–398.

[138] R. Seidel : Presentation at the AMS-IMS-SIAM Joint Summer Research Conference on *Discrete and Computational Geometry*, Santa Cruz, July 1986.

[139] I. Shemer : "Neighborly polytopes". *Israel Journ. Math.* **43** (1982) 291–314.

[140] J. Simutis : "Geometric realizations of toroidal maps". Ph.D. thesis, University of California at Davis, 1977.

[141] E. Steinitz, H. Rademacher : "Vorlesungen über die Theorie der Polyeder", Springer Verlag, Berlin, 1934; Reprinted by Springer, Berlin, 1976.

[142] E. Steinitz : "Konfigurationen der projektiven Geometrie". Encyklop. Math. Wiss. **3** (Geometrie) (1910) 481–516.

[143] G. Stengle : "A Nullstellensatz and a Positivstellensatz in semi-algebraic geometry". *Math. Ann.* **207** (1974) 87–97.

[144] B. Sturmfels : "Zur Realisierbarkeit orientierter Matroide", Diplomarbeit, TH Darmstadt, 1985.

[145] B. Sturmfels : "Boundary complexes of convex polytopes cannot be characterized locally", *Journ. London Math. Soc.* **35** (1987) 257–269.

[146] B. Sturmfels : "Some applications of affine Gale diagrams to convex polytopes with few vertices", *SIAM J. Discrete Math.* **1** (1988) 121-133.

[147] B. Sturmfels : "Oriented matroids and combinatorial convex geometry". Dissertation, Darmstadt 1987.

[148] B. Sturmfels : "Aspects of computational synthetic geometry, I: Algorithmic coordinatization of matroids", *Proceedings of Computer-aided Geometric Reasoning, INRIA, Antibes (France)*, 1987.

[149] B. Sturmfels : "Neighborly polytopes and oriented matroids", *European J. Combinatorics* **9** (1988) 537–546.

[150] B. Sturmfels : "Computing final polynomials and final syzygies using Buchberger's Gröbner basis method", *Results in Math.*,**15** (1989) 551–360.

[151] B. Sturmfels : "Simplicial polytopes without the isotopy property", I.M.A. Preprint Series # 410, University of Minnesota, April 1988.

[152] B. Sturmfels, N. White : "Rational realization of 11_3- and 12_3-configurations" in H. Crapo et. al.: *Symbolic Computations in Geometry*, I.M.A. Preprint, University of Minnesota, 1988.

[153] P. Suvorov : "Isotopic but not rigidly isotopic plane systems of straight lines", in O.Y. Viro (ed.): *Topology and Geometry – Rohlin Seminar*, Lecture Notes in Mathematics **1346**, 545–556, Springer, Heidelberg, 1988.

[154] A. Tarski, "A Decision Method for Elementary Algebra and Geometry", 2nd revised ed., Univ. of California Press, 1951.

[155] K. Truemper : "On the efficiency of representability tests for matroids", Europ. J. Combinatorics **3** (1982) 275–291.

[156] P. Vamos : "The missing axiom of matroid theory is lost forever". *Journ. London Math. Soc.* (2),**18** (1978) 403–408.

[157] A.M. Vershik : "Topology of the convex polytopes' manifolds, the manifold of the projective configurations of a given combinatorial type and representations of lattices" in O.Y. Viro (ed.): *Topology and Geometry – Rohlin Seminar*, Lecture Notes in Mathematics **1346**, 557–581, Springer, Heidelberg, 1988.

[158] D.J.A. Welsh : "Matroid theory", Academic Press, London, 1976.

[159] H. Weyl : "The classical groups - their invariants and representations". Princeton University Press, 1939.

[160] N. White : "The bracket ring of a combinatorial geometry. I", *Trans. Amer. Math. Soc.* **202** (1975) 79–103.

[161] N. White (Ed.) : "Theory of Matroids", Encyclopedia of Math. **26**, Cambridge University Press, 1986.

[162] N. White : "Combinatorial Geometries", Chapter I : "Coordinatizations", Cambridge University Press, 1987.

[163] N. White : "The transcendence degree of a coordinatization of a combinatorial geometry". *J. Combinatorial Theory* **B 29** (1980) 168–175.

[164] N. White : "A non-uniform oriented matroid which violates the isotopy property", *Discrete Comput. Geometry* 4(1989) 1–2.

Index

Vol. 1201: Curvature and Topology of Riemannian Manifolds. Proceedings, 1985. Edited by K. Shiohama, T. Sakai and T. Sunada. VII, 336 pages. 1986.

Vol. 1202: A. Dür, Möbius Functions, Incidence Algebras and Power Series Representations. XI, 134 pages. 1986.

Vol. 1203: Stochastic Processes and Their Applications. Proceedings, 1985. Edited by K. Itô and T. Hida. VI, 222 pages. 1986.

Vol. 1204: Séminaire de Probabilités XX, 1984/85. Proceedings. Edité par J. Azéma et M. Yor. V, 639 pages. 1986.

Vol. 1205: B.Z. Moroz, Analytic Arithmetic in Algebraic Number Fields. VII, 177 pages. 1986.

Vol. 1206: Probability and Analysis, Varenna (Como) 1985. Seminar. Edited by G. Letta and M. Pratelli. VIII, 280 pages. 1986.

Vol. 1207: P.H. Bérard, Spectral Geometry: Direct and Inverse Problems. With an Appendix by G. Besson. XIII, 272 pages. 1986.

Vol. 1208: S. Kaijser, J.W. Pelletier, Interpolation Functors and Duality. IV, 167 pages. 1986.

Vol. 1209: Differential Geometry, Peñíscola 1985. Proceedings. Edited by A.M. Naveira, A. Ferrández and F. Mascaró. VIII, 306 pages. 1986.

Vol. 1210: Probability Measures on Groups VIII. Proceedings, 1985. Edited by H. Heyer. X, 386 pages. 1986.

Vol. 1211: M.B. Sevryuk, Reversible Systems. V, 319 pages. 1986.

Vol. 1212: Stochastic Spatial Processes. Proceedings, 1984. Edited by P. Tautu. VIII, 311 pages. 1986.

Vol. 1213: L.G. Lewis, Jr., J.P. May, M. Steinberger, Equivariant Stable Homotopy Theory. IX, 538 pages. 1986.

Vol. 1214: Global Analysis – Studies and Applications II. Edited by Yu.G. Borisovich and Yu.E. Gliklikh. V, 275 pages. 1986.

Vol. 1215: Lectures in Probability and Statistics. Edited by G. del Pino and R. Rebolledo. V, 491 pages. 1986.

Vol. 1216: J. Kogan, Bifurcation of Extremals in Optimal Control. VIII, 106 pages. 1986.

Vol. 1217: Transformation Groups. Proceedings, 1985. Edited by S. Jackowski and K. Pawalowski. X, 396 pages. 1986.

Vol. 1218: Schrödinger Operators, Aarhus 1985. Seminar. Edited by E. Balslev. V, 222 pages. 1986.

Vol. 1219: R. Weissauer, Stabile Modulformen und Eisensteinreihen. III, 147 Seiten. 1986.

Vol. 1220: Séminaire d'Algèbre Paul Dubreil et Marie-Paule Malliavin. Proceedings, 1985. Edité par M.-P. Malliavin. IV, 200 pages. 1986.

Vol. 1221: Probability and Banach Spaces. Proceedings, 1985. Edited by J. Bastero and M. San Miguel. XI, 222 pages. 1986.

Vol. 1222: A. Katok, J.-M. Strelcyn, with the collaboration of F. Ledrappier and F. Przytycki, Invariant Manifolds, Entropy and Billiards; Smooth Maps with Singularities. VIII, 283 pages. 1986.

Vol. 1223: Differential Equations in Banach Spaces. Proceedings, 1985. Edited by A. Favini and E. Obrecht. VIII, 299 pages. 1986.

Vol. 1224: Nonlinear Diffusion Problems, Montecatini Terme 1985. Seminar. Edited by A. Fasano and M. Primicerio. VIII, 188 pages. 1986.

Vol. 1225: Inverse Problems, Montecatini Terme 1986. Seminar. Edited by G. Talenti. VIII, 204 pages. 1986.

Vol. 1226: A. Buium, Differential Function Fields and Moduli of Algebraic Varieties. IX, 146 pages. 1986.

Vol. 1227: H. Helson, The Spectral Theorem. VI, 104 pages. 1986.

Vol. 1228: Multigrid Methods II. Proceedings, 1985. Edited by W. Hackbusch and U. Trottenberg. VI, 336 pages. 1986.

Vol. 1229: O. Bratteli, Derivations, Dissipations and Group Actions on C*-algebras. IV, 277 pages. 1986.

Vol. 1230: Numerical Analysis. Proceedings, 1984. Edited by J.-P. Hennart. X, 234 pages. 1986.

Vol. 1231: E.-U. Gekeler, Drinfeld Modular Curves. XIV, 107 pages. 1986.

Vol. 1232: P.C. Schuur, Asymptotic Analysis of Soliton Problems. VIII, 180 pages. 1986.

Vol. 1233: Stability Problems for Stochastic Models. Proceedings, 1985. Edited by V.V. Kalashnikov, B. Penkov and V.M. Zolotarev. VI, 223 pages. 1986.

Vol. 1234: Combinatoire énumérative. Proceedings, 1985. Edité par G. Labelle et P. Leroux. XIV, 387 pages. 1986.

Vol. 1235: Séminaire de Théorie du Potentiel, Paris, No. 8. Directeurs: M. Brelot, G. Choquet et J. Deny. Rédacteurs: F. Hirsch et G. Mokobodzki. III, 209 pages. 1987.

Vol. 1236: Stochastic Partial Differential Equations and Applications. Proceedings, 1985. Edited by G. Da Prato and L. Tubaro. V, 257 pages. 1987.

Vol. 1237: Rational Approximation and its Applications in Mathematics and Physics. Proceedings, 1985. Edited by J. Gilewicz, M. Pindor and W. Siemaszko. XII, 350 pages. 1987.

Vol. 1238: M. Holz, K.-P. Podewski and K. Steffens, Injective Choice Functions. VI, 183 pages. 1987.

Vol. 1239: P. Vojta, Diophantine Approximations and Value Distribution Theory. X, 132 pages. 1987.

Vol. 1240: Number Theory, New York 1984–85. Seminar. Edited by D.V. Chudnovsky, G.V. Chudnovsky, H. Cohn and M.B. Nathanson. V, 324 pages. 1987.

Vol. 1241: L. Gårding, Singularities in Linear Wave Propagation. III, 125 pages. 1987.

Vol. 1242: Functional Analysis II, with Contributions by J. Hoffmann-Jørgensen et al. Edited by S. Kurepa, H. Kraljević and D. Butković. VII, 432 pages. 1987.

Vol. 1243: Non Commutative Harmonic Analysis and Lie Groups. Proceedings, 1985. Edited by J. Carmona, P. Delorme and M. Vergne. V, 309 pages. 1987.

Vol. 1244: W. Müller, Manifolds with Cusps of Rank One. XI, 158 pages. 1987.

Vol. 1245: S. Rallis, L-Functions and the Oscillator Representation. XVI, 239 pages. 1987.

Vol. 1246: Hodge Theory. Proceedings, 1985. Edited by E. Cattani, F. Guillén, A. Kaplan and F. Puerta. VII, 175 pages. 1987.

Vol. 1247: Séminaire de Probabilités XXI. Proceedings. Edité par J. Azéma, P.A. Meyer et M. Yor. IV, 579 pages. 1987.

Vol. 1248: Nonlinear Semigroups, Partial Differential Equations and Attractors. Proceedings, 1985. Edited by T.L. Gill and W.W. Zachary. IX, 185 pages. 1987.

Vol. 1249: I. van den Berg, Nonstandard Asymptotic Analysis. IX, 187 pages. 1987.

Vol. 1250: Stochastic Processes – Mathematics and Physics II. Proceedings 1985. Edited by S. Albeverio, Ph. Blanchard and L. Streit. VI, 359 pages. 1987.

Vol. 1251: Differential Geometric Methods in Mathematical Physics. Proceedings, 1985. Edited by P.L. García and A. Pérez-Rendón. VII, 300 pages. 1987.

Vol. 1252: T. Kaise, Représentations de Weil et GL_2 Algèbres de division et GL_n. VII, 203 pages. 1987.

Vol. 1253: J. Fischer, An Approach to the Selberg Trace Formula via the Selberg Zeta-Function. III, 184 pages. 1987.

Vol. 1254: S. Gelbart, I. Piatetski-Shapiro, S. Rallis. Explicit Constructions of Automorphic L-Functions. VI, 152 pages. 1987.

Vol. 1255: Differential Geometry and Differential Equations. Proceedings, 1985. Edited by C. Gu, M. Berger and R.L. Bryant. XII, 243 pages. 1987.

Vol. 1256: Pseudo-Differential Operators. Proceedings, 1986. Edited by H.O. Cordes, B. Gramsch and H. Widom. X, 479 pages. 1987.

Vol. 1257: X. Wang, On the C*-Algebras of Foliations in the Plane. V, 165 pages. 1987.

Vol. 1258: J. Weidmann, Spectral Theory of Ordinary Differential Operators. VI, 303 pages. 1987.